农民教育培训农业农村部"十三五"规划教材

Practical English Course for Farmers

农民实用英语教程

姚 岚 唐 帅 主编

中国农业出版社
北 京

编写人员名单

主　编 姚　岚　唐　帅
副主编 卢　鹿　高　山　张　飒
参　编 李　龙　任志萍

前言

随着全球经济一体化进程的加快,农业的国际化趋势越来越明显。越来越多的高素质农民已经不满足于简单的耕种,而是要把农业当作国际化背景下的产业,以广阔的心胸放眼世界。《农民实用英语教程》就是针对高素质农民的需求,旨在满足这一群体对于农业发展的更高追求,为高素质农民放眼世界提供助力。

一、编写依据

本教材以面向农业国际化的高素质农民英语教学内容、课程体系和教学方法改革为出发点,充分考虑农民学员的英语实用需求,通过对目标群体的充分调研,以全面提高学员的英语运用能力、交际能力为目标,力争在博采众长的基础上自成一家,提升学员的综合分析能力和跨文化交际能力。

二、创新理念

1. 宏观与微观的统一

本教材在内容安排上采用了先总后分的原则,先总体讨论了农业的现状,再对目前高素质农民所关注的具体热点问题进行了专题讲解。既有高屋建瓴的宏阔,又有一叶知秋的细微,合为有机整体,分而独立成篇,达到了宏观与微观的和谐统一。

2. 英语与专业的统一

本教材对我国农业的现状和相关问题进行了追本溯源的介绍,为学员了解相关知识提供了可行性参考,是英语专业性和农业专业性的完美结合。

3. 深入与浅出的统一

本教材既兼顾了英语与专业，又照顾到了学员的英语水平，遵循"辞达"的原则，体现出内容深入而词句浅出的特性。

4. 学习与应用的统一

学习的目的在于应用，而非记忆。一本书并不能满足读者所有的需求，因此本教材大胆尝试摆脱单纯的教学，通过以用带学的方式，激发学员自觉探求知识的热情。

5. 教育与娱乐的统一

语言学习不应该是枯燥无味的，而应是充满乐趣的，让人在潜移默化中掌握知识。本教材力图避免高高在上的说教式教学，在每个章节后面都设有一个娱乐版块，寓教于乐，使学员能够轻松地掌握一些有趣的英语知识，对自己的英语水平也能有一个很好的自测和提升。

三、教材构成

《农民实用英语教程》共分为八个单元，每单元由如下部分组成：

Lead-in：这是课程的导入部分，以问题的提出来激发学员的英语学习兴趣，并引发学员的思考，在新旧知识间构筑有机联系。

Intensive Reading：这是课程的主体部分，在内容上是对导入部分所提出问题的探讨，旨在实现对学员的专业知识和英语能力的双重培养。

Extended Reading：此部分是对相关文化的简介，是对主体部分的有益补充。如果说 Intensive Reading 是授之以"鱼"，那么这部分就是授之以"渔"。

Exercises：这是课程的巩固部分。以任务为驱动，培养学员对语言材料的掌握能力。

Writing：此部分侧重介绍各类应用文体写作策略，以提高学员的涉外业务实际写作能力。

Study for Fun：此部分为了让农民学员体验英语学习的快乐，放松一刻。

前言

《农民实用英语教程》由沈阳农业大学、中国农业科学院研究生院、沈阳大学、辽宁农业职业技术学院等单位的多位教授与英语教学专家分工写作、集体编写而成。沈阳农业大学的毕凤春教授、段玉玺教授和中国医科大学的王焱教授担任了本教材的审定与顾问工作。

本教材在策划、设计与编写过程中得到了中国农业出版社的大力支持，沈阳农业大学各学院专业教师为本教材提供原版论文予以赞助，外语教学部以及沈阳农业大学继续教育学院的领导为本教材的编写和使用提供了鼎力支持，再此一并表示感谢。

本教程在编写过程中，参考了许多农业专业英语相关论著，吸收了许多专家同仁的观点和例句，但为了行文方便，未一一注明。在此，特向在本教材中引用和参考的教材、专著、文章的作者表示诚挚的谢意。

由于编者能力及经验所限，本教材中难免有一些不尽如人意之处，敬请专家读者批评指正。

编　者

2022 年 12 月

前言

Unit One　　Agricultural Problems ………………………………………… 1

Unit Two　　Artificial Intelligence in the Agriculture Industry ………… 14

Unit Three　　Agriculture Outlook ………………………………………… 26

Unit Four　　Ecological Agriculture ……………………………………… 37

Unit Five　　Agricultural Product Processing …………………………… 49

Unit Six　　Plant Protection ………………………………………………… 60

Unit Seven　　Animal Husbandry and Veterinary ……………………… 71

Unit Eight　　Organic Food ………………………………………………… 84

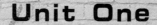

Unit One

Agricultural Problems

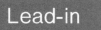

Discussion: What are the Causes of Agricultural Problems?

- Increasing world population.
- Economic development in developing countries.
- Meat-eating dietary habit.
- Farmers' demand for cereal crops.
- Gaps between the rich and the poor, developed and developing countries.
- Food security problems.
- Food processing problems.
- Environmental problems.
- Energy and economic problems.
- Disasters such as drought, flood, wars, and famine.

Agricultural Problems

Grain is a commodity of **strategic** significance. The basic self-sufficiency of **cereal** grains is vital to the livelihood of individuals and the national security of China, a country with a population of 1.4 billion. Take Northeast China as an example. In China, the Northeast is the most **abundant** region in agricultural resources, with the grain industry growing faster than other regions of China. Northeast China has the greatest potential for development and contributes the most to China's national food security. Northeast China is presently an important production area of commodity grain and **livestock** products.

Northeast China is rich in agricultural resources. Its vast ranges of flat and **fertile** farmland are suitable for **mechanized** agricultural production. Moreover, with a low rate of pest occurrence because of the cool climate, agricultural production in this region requires relatively low amounts of fertilizers and pesticides. This region is one of China's major production areas for crop and livestock products (**corn**, rice, **soybean**, pork, beef, mutton, etc.). Nevertheless, there is currently a concern for both the regional food security of Northeast China and the national food security of China.

Overall, Northeast China has become one of China's major areas for the production of grains and the rearing of livestock. During an inspection of Northeast China in 2018, President Xi Jinping **emphasized** that the region was one of China's major industrial and agricultural areas and played an extremely important strategic role in **safeguarding** China's national defense security, food security, ecological security, energy security and industry security. **Thereby**, this region is recognized to be vital to China's national development and **coordination**.

Nevertheless, as Northeast China has an unbalanced crop production structure, particularly **based** on the ratio between foods, industries, and **forage** crops, the region is currently suffering from structural and short-term **surpluses** of some agricultural products. Furthermore, there is a lack of water **conservancy facilities**, and **existing** facilities are poorly equipped and outdated, causing a low rate of water resource **utilization** and excessive **exploitation** of groundwater. Currently, this is the major **barrier** to Northeast China's **sustainable** food security.

To improve the agricultural **infrastructure** and safeguard the supply of major agricultural products, policy makers should focus on: establishing and improving mechanisms for **subsidizing** major grain-producing areas and increasing the subsidies for major grain-producing counties; **strengthening** high-standard farmland construction and black-soil conservation; strengthening the

Unit One
Agricultural Problems

construction and **renovation** of water-saving facilities in medium and large **irrigation** areas to improve the management of water resources; and increasing the support for technological innovation in agricultural crop production, full mechanization, innovative agriculture and green agriculture. By **establishing** these goals, a food crop production strategy based on farmland management and technological **application** will be **implemented** to safeguard sustainable food security and agricultural development.

Word Bank

strategic [strə'tiːdʒik]	a.	relating to strategy 战略上的；策略的
cereal ['siəriəl]	n.	any kind of grain used for food 谷物；谷类
	a.	relating to grain 谷类的；谷类制成的
abundant [ə'bʌndənt]	a.	present in great quantity 丰富的；充裕的
livestock ['laivstɔk]	n.	animals kept for use or profit 牲畜；家畜
fertile ['fəːtail]	a.	capable of reproducing 肥沃的；能生育的
mechanized ['mekənaizd]	a.	equipped with machinery 机械化的
corn [kɔːn]	n.	maize（美）玉米；（英）谷物
soybean ['sɔibiːn]	n.	bean grown as food or for oil 大豆；黄豆
emphasize ['emfəsaiz]	v.	stress or single out as important 强调；着重
safeguard ['seifgaːd]	v.	make safe 保护；护卫
	n.	measures to prevent harms 保护；预防措施
thereby [ˌðɛə'bai]	ad.	by that means or because of that 从而；因此
coordination [kəuˌɔːdi'neiʃn]	n.	the act of putting into proper relation 协调；配合

base [beis]	v.	use as a basis for 根据；建于……之上
	n.	lowest part of anything 基底；基础
forage ['fɔridʒ]	n.	animal food for browsing or grazing 饲料
	v.	collect or look around for food 觅食
surplus ['sə:pləs]	n.	a quantity much larger than needed 剩余；过剩
	a.	more than needed or required 剩余的；过剩的
conservancy [kən'sə:vənsi]	n.	official conservation 管理；保护
facility [fə'siliti]	n.	aids which make it easy to do things 设施；设备
exist [ig'zist]	v.	have an existence, be extant 存在；生存
utilization [ju:təlai'zeiʃn]	n.	being used for a particular purpose 利用；使用
exploitation [,eksplɔi'teiʃn]	n.	the use of natural resources 利用；开发
barrier ['bæriə(r)]	n.	hindrance or prevention 障碍物；屏障
sustainable [sə'steinəbl]	a.	capable of being sustained 可持续的；足可支撑的
infrastructure ['infrəstrʌktʃə]	n.	the basic structure or features of a system or organization 基础设施；公共建设
subsidize ['sʌbsidaiz]	v.	give granted benefits or money to 资助；补贴
strengthen ['streŋθn]	v.	make or become stronger 加强；加固
renovation [,renə'veiʃn]	n.	the act of improving by renewing 革新；活力
irrigation [,iri'geiʃn]	n.	supplying the dry land with water 灌溉；冲洗

establish [iˈstæbliʃ]	v.	start or creat 建立；设立
application [ˌæpliˈkeiʃn]	n.	put into practical use 应用；申请
implement [ˈimplimənt]	v.	apply with purposes or designs 实施；使生效
	n.	instrumentation used to effect an end 工具；手段

Extended Reading

Festival in China (1)

1. Spring Festival（春节）: It is the most unique and important traditional Chinese holiday. It marks the beginning of the Chinese lunar New Year. The Spring Festival is a time for family reunion. People working away from home will travel a long distance to go back to join their families. On New Year's Eve, the whole family gets together, making *jiaozi*（饺子）and enjoying the Eve dinner. Every household will paste up Spring Festival couplets and Chinese New Year pictures. Around 24 o'clock at the New Year's Eve, many families will set off fireworks and firecrackers to greet the new days and send off the old ones, hoping to cast away bad luck and bring forth good luck.

2. Qingming Festival（清明节）: It's around April 5th, and is a time for people to sweep the tombs of departed ones. People put food, flowers, tea, wine, chopsticks in front of the memorial tablet, burn incense and bow. People observe the day by honoring their ancestors.

3. Duanwu Festival（端午节）: It's on the 5th day of the 5th lunar month of a year, and is also known as Dragon Boat Festival. It originated in China and used to serve as a time to ward off illness and prevent epidemics. It is said to honor the brilliant patriotic poet Qu Yuan. The main activities of the festival include eating *zongzi*（粽子）, holding dragon boat races, hanging herbal medicine and drinking yellow rice wine.

4. National Day(国庆节): October 1st is the National Day of the People's Republic of China (PRC), which is a public holiday for the whole country. It is an important day that marks the founding of the PRC. On that day, there

are many kinds of celebrations throughout the country, from the central government to the ordinary people. Public places, including big squares and parks are decorated with festive themes. It is also a good time to have a good rest as well as traveling.

5. Mid-Autumn Festival（中秋节）: It's on the 15th day of the 8th lunar month of a year and is the second largest traditional festival in China. It is a time for family members and loved ones to enjoy the full moon, wishing harmony and luck. Adults usually indulge in fragrant moon cakes of many varieties with a good cup of hot Chinese tea as well as juicy grapes, while the children run around with their brightly-lit lanterns.

Exercises

Ⅰ. Reading Comprehension

Directions: *Choose the best answer for the following questions.*

1. What is the solution suggested by the author to improve the agricultural infrastructure and safeguard the supply of major agricultural products? (　　　)

   A. Implementing a food crop production strategy based on farmland management and technological application.

   B. Policy makers should aim at establishing and improving mechanisms.

   C. Accelerating the structural reform of the agricultural supply-side and pursue well-coordinated production of grain, industries, and forage crops.

   D. Understanding the recent evolution of the food consumption structure.

2. Which of the following statements can be inferred as a threat to food-security? (　　　)

   A. Serious soil and water loss of black-soil farmland.

   B. Food consumption structures of rural and urban populations.

   C. Corn serves as the primary contributor to the increase in food output and its farming area has markedly increased.

   D. Cool climate requires relatively low amounts of fertilizers and pesticides for agricultural production.

3. What attitude can be inferred about the author towards China's future food security? (   )

 A. The issue of China's future food security is essentially the same wine in a blue bottle.

 B. China's future food security problems will not be overcome as quickly as that in the western countries because of its limitations.

 C. The author is quite sure that China will become a pioneer to develop food security strategies in the future.

 D. Timely personalized support is provided to create the agricultural revolution in China.

4. According to the text, how should China meet its needs of further development? (   )

 A. China should improve the water resource management in medium and large irrigation areas.

 B. China should meet its needs of the basic self-sufficiency of cereal grains.

 C. China should make its people aware of the coming shortage of food.

 D. China should equip itself with knowledge and advanced technologies to meet the need of the 21st century economy.

5. What is the main idea of the text? (   )

 A. The reasons why China has developed so quickly in the past 20 years.

 B. Northeast China's new agricultural revolution and its achievements.

 C. The problems to be solved regarding the construction and renovation of water-saving facilities in medium and large irrigation areas.

 D. China's current needs regarding economic booming and food security challenges.

II. Vocabulary

Section A

Directions: *Match the English words or phrases with their Chinese equivalents.*

1. (　　) natural resources　　　　A. 耕地资源
2. (　　) land resources　　　　　 B. 矿产资源
3. (　　) water resources　　　　　C. 自然资源
4. (　　) climate resources　　　　D. 土地资源
5. (　　) biological resources　　　E. 水资源
6. (　　) mineral resources　　　　F. 生物资源
7. (　　) farm resources　　　　　 G. 气候资源
8. (　　) fresh water resources　　 H. 淡水资源
9. (　　) power resources　　　　　I. 电力资源
10. (　　) food resources　　　　　J. 食品资源

Section B

Directions：*Translate the following words and phrases.*

1. 小麦
2. 谷物
3. 粮食
4. 耕地
5. 面粉
6. 加工
7. 价格
8. 玉米
9. 庄稼
10. 水稻
11. import
12. export
13. the World Bank
14. rise
15. drop
16. the World Financial Crisis
17. drought
18. food security
19. fertilizer
20. insecticide

Ⅲ. Situational Speaking

Asking for Directions

Section A

A：Excuse me. May I ask you some questions?
B：Sure.

A: Where is the post office?

B: It's next to the hospital.

A: Then, where is the hospital?

B: Turn left, and go straight for one minute. Then you can get there.

A: 劳驾，可以问一下吗?

B: 可以。

A: 邮局在哪里?

B: 在医院旁边。

A: 那医院在哪里?

B: 先左拐，再直行一分钟，就到了。

Section B

A: Excuse me. Where am I on this map?

B: We are here, former address of Qinghua Hot Spring. We are in the heart of the city.

A: Oh! I think I'm lost. Can I go from here to the present address of Qinghua Hot Spring?

B: Head straight up the street about two blocks then turn right.

A: Is it far?

B: No. It's only a fifteen-minute walk.

A: 劳驾，请问我在地图上的什么地方?

B: 我们在这里，清华池原址。我们现在在市中心。

A: 哦！我想我迷路了。我能否从这里到清华池现址呢?

B: 顺这条街一直走过两个街区，然后右转。

A: 远不远?

B: 不远。15分钟就走到了。

How to Write Business Contracts

A contract is the principal means used by a business firm to keep in touch with its customers. The quality of paper and an attractive letter-head play an important part in writing a contract, but they are less important than the message they carry. A contract does not need elegant language, but it does require us to express ourselves accurately in plain language that is clear and readily understood.

Rules of Writing a Contract

- Write naturally and sincerely.
- Avoid wordiness.
- Be courteous and considerate.
- Be precise.
- Plan your contract.
- Check your contract.

[Sample]

Wheat Purchase Business Contract

1. Both Parties（双方）

Buyer: China National Chemicals Import & Export Corporation, Beijing, China

Seller: Chemicals Export Corporation, Canada

The buyer agrees to buy and the seller agrees to sell the commodity on the following terms and conditions.

2. Name of Commodity and Quantity（商品名称和数量）

Wheat, one hundred thousand tons. Among it, there is 10% more or less at the buyer's option. The buyer shall inform the seller of this option within one month prior to the completion of this contract.

3. Specifications（规格）

Wheat flour

Moisture

Energy

Protein

Fat

Boarding Fibers

Carbohydrate

4. Price（价格）

Total value: US $1,200,000. (One Million Two Hundred Thousand US Dollars.)

5. Destination（目的地）

Dalian Port

6. Period of Delivery（交货期）

50,000 tons are to be delivered in October, 2004 at Vancouver Port.

25,000 tons are to be delivered from October to December, 2004 at Prince Rupert Port.

25,000 tons are to be delivered from January to March, 2005 at Prince Rupert Port.

7. Weighing（称重约定）

Wheat shall be weighed under the supervision of both the seller's and the buyer's representatives.

8. Inspection（检查）

The determination of quality of wheat is subject to the results of analysis of the representative samples drawn from the actually landed cargo, conducted by the China Commodity Inspection Bureau after arrival of the goods at destination. The buyer shall have the right to claim against the seller for the compensation of losses within 60 days after the arrival of the goods at the port of destination.

9. Invoicing（发票）

Should the contents of wheat fall below the guaranteed, an allowance on the basis of 2% of the sale price shall be calculated and deducted from the purchase price. This will be made out in the invoice.

10. Payment（支付）

The buyer shall open accounts in the Bank of China, Beijing. The buyer should open the relative letter of credit 20 days before the arrival of the carrying vessel at the port of loading with validity for 90 days from the date of opening.

11. Terms of Shipment（运输条款）

Insurance will be covered by the buyer. The buyer shall undertake to charter the carrying vessel. The seller shall confirm upon receipt of the above telegram. The buyer' chartering agent shall make direct contact with the seller from time to time.

12. Irresistible Forces（不可抗力）

The seller shall not be responsible for late delivery of the goods owing to the generally recognized "irresistible forces" causes. However, in such case, the seller shall telegraph the buyer immediately and deliver in 14 days to the buyer a certificate of the occurrence issued by the Government Authorities or the Chamber of Commerce at the place where the accident occurs as evidence.

13. Arbitration（仲裁）

All disputes in connection with this contract shall be settled amicably by negotiation. In case no settlement can be reached, the case under dispute may then be submitted for arbitration.

14. Drafting Date（合同起草日期）

On 25th July, 2003

15. Drafting Place（合同起草地点）

Beijing, China

Assignment: *Write a simulated contract of rice.*

_____
_____
_____
_____
_____
_____

Study for Fun

Relax and Enjoy the Song

Directions: *Some words are missing in the song. While you are listening, please fill in the missing words.*

Beautiful Island

Last night I dreamt of San Pedro
Just like I'd never gone, _____
A young girl with eyes like the desert

Unit One

Agricultural Problems

It all seems like yesterday, not far away
Tropical the island _____
All of nature, wild and _____
This is where I love to be
And when the samba played
The sun would set so high
Ring through my ears and sting my eyes
You Spanish lullaby
I fell in love with San Pedro
Warm wind carried on the sea, he called to me
I _____ that the days would last
They went so fast
I want to be where the sun warms the _____
When it's time for siesta you can watch them _____
Beautiful faces, no _____ in this world
Where a girl loves a boy
And _____

Unit Two

Artificial Intelligence in the Agriculture Industry

Lead-in

Discussion: What is the Future Tendency of AI Use in China's Agriculture?

- Improve the intelligent level of agricultural machinery equipment.
- Promote the transformation and upgrading of China's agricultural machinery and equipment, and agricultural mechanization as well.
- Accelerate the research, development, promotion and application of agricultural machinery equipment in hilly areas and in the production of fruit, vegetable, tea, livestock, poultry and aquaculture.
- Promote the integration of agricultural machinery techniques and crop cultivation techniques.
- Encourage Internet enterprises to establish AI agricultural service platforms linking production and marketing, and to strengthen agricultural information monitoring and predicting systems.
- Develop AI digital agriculture, remote sensing, and the "Internet +" modern agriculture initiative.
- Develop AI digital meteorology to serve agriculture.
- Help to implement agricultural projects.

Unit Two

Artificial Intelligence in the Agriculture Industry

Intensive Reading

Automatic Visual Grading System for Grafting Seedlings

Grafting is a technique whereby **branches** or **buds** from one plant are **inserted** into the appropriate parts of another plant that has a strong **affinity**. A branch or bud to be cultivated is encouraged to **fuse** or graft onto a strong vital plant so that the **scion** can grow, **blossom**, and **bear** fruit using rich soil. Grafting technique can reduce soil-borne diseases, enhance stress resistance, increase **yield** and improve fruit quality. Currently, the grafting technology is rapidly developing over the world in order to adapt to the needs of agricultural production. At the same time, grafting robot technology has been developed. Some semi-automatic grafting robots have been developed and **commercialized** for cucumbers, tomatoes, and watermelons in Japan, South Korea, the Netherlands and China. In these countries, the working **efficiency** of semi-automatic grafting robots is three times higher than the traditional **manual** grafting. Furthermore, for most automatic grafting machines, the grafting speed can reach up to 600 seedlings per hour. However, since the machines still require **auxiliary** manual work, they can only achieve a semi-automatic **status**. The scion and stock plants must be **classified** so that the same size of seedlings can be placed **properly** in contact with each other in the **process** of grafting. This method can reduce the loss of tissue **fluid**, and the survival rate of the grafted seedlings can also be **guaranteed.**

Machine vision is widely used in many fields, such as industry, animal **husbandry**, fisheries and **horticulture.** In recent years, the applications of machine vision have been deeply studied in various fields, for instance, in grading of **panel** surfaces, high temperature measurements of surface properties, automatic grading systems for fresh fruits, vegetables, areca nuts, eggs and other food products. Moreover, researchers have recently presented some **algorithms** for processing images, which is the core of machine vision technology.

The main algorithm includes color space **conversion**, gray-scale **transformation**, image **binarization**, noise smoothing and **morphological operations.** The machine vision system of automatic grafting machine still has many **shortcomings**, and many problems like lower **accuracy** remain to be solved. Different algorithms of image processing technologies have been adopted in many fields, for example, supporting **vector** machines. Although these algorithms can **distinguish** the characteristics of measured objects in detail, the Calculation process is too complex to be suitable for grading the characteristics of grafting seedlings. Furthermore, because of the variety of seedling, the algorithms are inaccurate in grading seedlings for some machine vision systems. Comparison of natural and **artificial** light source has not been reported, either. For these reasons, it is important and necessary to develop simple and **practical** visual grading systems for **muskmelon** grafting machines.

Currently, semi-automatic grafting machines have been developed, while the fully automatic grafting machines remain in development process. A fully automatic grafting machine is supposed to have higher grafting efficiency than semi-automatic machines. Developing an automatic visual grading system of vegetable grafting machines for seedlings grafting is **indispensable.** The main research objectives are as follows: ① to design an automatic visual grading system which will include image **acquisition** module, image processing module, and control module; ② to choose **appropriate** light sources in the image acquisition module.

The experimental equipment includes a light source, PC computer and an image processing kit model. A blue light-emitting **diode** (LED) panel **backlighting** was used as an illumination source with three strength grades—high, medium, and low. A comparative analysis of a vegetable grafting machine and machine vision technology in agricultural production was conducted. It is identified that the blue light illumination can be used to capture the best images, based on machine vision and image process theory.

Word Bank

graft [grɑːft]	v.	cut a piece form a living plant and attach it to another plant 嫁接
branch [bræntʃ]	n.	a division of some larger organization 分支；分部
	v.	grow and send out branches 发出新枝；岔开

Unit Two

Artificial Intelligence in the Agriculture Industry

bud [bʌd]	n.	a compact knoblike growth on a plant 芽；花蕾
	v.	(of a plant) form a bud 萌芽；发芽
insert [in'sə:t]	v.	put or introduce into sth. 插入；置入
	n.	a folded section placed into another 插页；插物
affinity [ə'finəti]	n.	a spontaneous or natural liking 姻亲关系；近似
fuse [fju:z]	v.	mix different elements together 融合；熔接
scion ['saiən]	n.	a young shoot or twig of a plant 幼芽；嫩芽
blossom ['blɔsəm]	v.	produce or yield flowers 开花；兴旺
	n.	reproductive flowers on a bush 花；开花期
bear [bɛə(r)]	v.	cause to be born 结果实
yield [ji:ld]	n.	production of a certain amount 产量；利润
	v.	produce or force 出产；产出（效果、收益等）
commercialize [kə'mə:ʃəlaiz]	v.	in a way for making profits 使商品化；靠……赚钱
efficiency [i'fiʃnsi]	n.	the ratio of the output of any system 效率；效能
manual ['mænjuəl]	a.	of or relating to the hands 体力的；手工的
	n.	a small handbook 说明书；小册子
auxiliary [ɔ:g'ziliəri]	a.	supplementary or additional 辅助的；备用的
status ['steitəs]	n.	the relative position or state of things 地位；状态

classify ['klæsifai]	v.	arrange or order by classes or categories 把……分类
properly ['prɔpəli]	ad.	in the right manner 适当地；正确地
process [prə'ses]	n.	a course of action to achieve a result 过程；进展
fluid ['fluːid]	n.	a liquid that has no fixed shape 流体；液体
	a.	(of a liquid) able to flow easily 流动的；不固定的
guaranteed [,gærən'tiːd]	a.	secured by written agreement 肯定的；保证的
husbandry ['hʌzbəndri]	n.	farming, especially when done carefully and well（尤指精心经营的）农牧业
horticulture ['hɔːtikʌltʃə]	n.	garden cultivation and operation 园艺；园艺学
panel ['pænl]	n.	separate part of a surface 仪表板；嵌板
algorithm ['ælgəriðəm]	n.	a process of calculations 算法；运算法则
conversion [kən'vəːʒn]	n.	an event that causes a transformation 转换；变换
transformation [trænsfə'meiʃn]	n.	a qualitative change 转化；转换
binarization [bainərai'zeiʃn]	n.	the process of converting to binary 二值化
morphological [,mɔːfə'lɔdʒikl]	a.	relating to the morphology 形态学的；形态的
operation [,ɔːpə'reiʃn]	n.	a business run on a large scale 操作；经营
shortcoming ['ʃɔːtkʌmiŋ]	n.	a failing or deficiency 缺点；短处
accuracy ['ækjərəsi]	n.	the quality of being near to the true value 精确度；准确性

Unit Two

Artificial Intelligence in the Agriculture Industry

vector ['vektə(r)]	n.	a variable quantity that can be resolved into components 矢量
distinguish [di'stiŋgwiʃ]	v.	mark as different 区分；辨别
artificial [ˌɑːti'fiʃəl]	a.	made by art rather than nature 人造的；仿造的
practical ['præktikl]	a.	concerned with actual use 实际的；实用性的
muskmelon ['mʌskˌmelən]	n.	a fruit with edible flesh and smell 甜瓜；香瓜
indispensable [ˌindi'spensəbl]	a.	absolutely necessary 必不可少的；必需的
acquisition [ˌækwi'ziʃn]	n.	an asset or object bought or obtained 获得；收获
appropriate [ə'prəupriət]	a.	suitable or proper 适当的；恰当的
diode ['daiəud]	n.	a thermionic tube having two electrodes 二极管
backlighting [bæk'laitiŋ]	n.	lighting from behind 逆光；倒光

Extended Reading

Festivals in China (2)

1. New Year's Day (元旦): On the first day of January, people celebrate the beginning of the Gregorian calendar year. Festivities include counting down to midnight (12:00 p.m.) on the night of New Year's Eve.

2. Lantern Festival (元宵节): The Lantern Festival is also called the Shangyuan Festival. It is a traditional Chinese festival. It is held on the 15th day of the first month of the lunar calendar every year.

3. The Double Ninth Festival (重阳节): The Double Ninth Festival, the 9th day of the 9th month of the Chinese lunar calendar, is a traditional festival of the Chinese nation. In ancient times, there were customs of climbing high to pray for blessings, appreciating chrysanthemums in autumn, wearing dogwood, offering sacrifices to gods and ancestors, and feasting for

longevity.

4. Qixi Festival（七夕节）: Qixi Festival (or "Qiqiao Festival", "The Begging Festival") falls on the 7th day of the 7th month of the lunar calendar (August according to the Gregorian calendar). This is a day devoted to romance, also known as Chinese Valentine's Day. It is also an important day for girls. In the evening, girls prepare melons and fruits to pray for a good marriage.

5. Laba Festival（腊八节）: Laba Festival is a traditional Chinese holiday celebrated on the 8th day of the 12th lunar month. It's a day for celebrating harvest, offering sacrifices to ancestors, and worshiping gods. But above all, it is a day for cooking Laba Congee（腊八粥）, a hodgepodge of grains, beans and dried fruits used to invoke peace and good fortune.

Exercises

Ⅰ. Reading Comprehension

Directions: *Choose the best answer for the following questions.*

1. Grafting technique cannot _____ . (   )
   A. ruin the plant quality
   B. reduce soil-borne diseases
   C. increase yield and fruit quality
   D. enhance stress resistance

2. Which of the following statements is true? (   )
   A. The grafting technology is widely used around the world.
   B. Some semi-automatic grafting robots will be commercialized.
   C. The grafting speed of all the automatic grafting machines can reach up to 600 seedlings per hour.
   D. The scion and stock plants must be attached.

3. What is the core of machine vision? (   )
   A. Automatic grading system.
   B. The field of high temperature measurements of surface properties.
   C. Some algorithms for processing images.

D. The field of grading of panel surfaces.

4. The main objective to develop an automatic visual grading system of vegetable grafting machine for seedling grafting is _____ . (   )

    A. to decline an automatic visual grading system

    B. to develop an automatic visual grading system

    C. to recycle appropriate light sources

    D. to give up appropriate light sources

5. In order to develop the fully automatic grafting machines, the experimental equipment includes _____ . (   )

    A. an image processing kit

    B. the internet

    C. automatic equipment

    D. a light source

Ⅱ. Vocabulary

Section A

Directions: *Match the English words or phrases with their Chinese equivalents.*

1. ( )	agricultural machinery equipment	A. 转型升级
2. ( )	improve the intelligent level	B. "互联网+" 现代农业
3. ( )	modern agricultural management system	C. 现代农业经营体系
4. ( )	a new agricultural pattern of opening up	D. 人工智能数字气象
5. ( )	linking production and marketing	E. 农机配套
6. ( )	"Internet +" modern agriculture	F. 提高智能化水平
7. ( )	AI agricultural service platforms	G. 人工智能农业服务平台
8. ( )	AI digital meteorology	H. 农业对外开放新格局
9. ( )	the transformation and upgrading	I. 农机装备
10. ( )	agricultural machinery integration	J. 产销衔接

Section B

Directions: *Translate the following words and phrases.*

1. 嫁接
2. 半自动
3. 机器人
4. 枝条
5. 组织液
6. 图像二值化
7. 甜瓜嫁接机
8. 幼苗多样性
9. 准确率
10. 农业生产
11. stress resistance
12. insert into
13. manual grafting
14. grafting survival rate
15. visual grading system
16. color space conversion
17. an image processing kit model
18. automatic grafting machine
19. image acquisition module
20. natural and artificial light source

Ⅲ. Situational Speaking
Ordering Foods

A：Sit down, madam, please! Here is the menu!

B：Thank you!

A：May I take your order?

B：Yes, of course.

A：What would you like?

B：Well, I will take this one and that one. (Pointing to the menu.) I would like a steak, some fries and green salad.

A：How would you like your steak? Rare, medium or well-done?

B：Medium, please. May I see the wine list, please?

A：Here you are.

B：And a bottle of wine, please.

A：Anything else, madam?

B：Fruit juice, please! That's all.

A：OK. Wait a moment!

B：Thank you!

A：女士，请坐！给您菜单！

B：谢谢！

A：可以点餐了吗?

B：是的，当然。

A：您想来点什么呢?

B：我想点这个和那个。(手指菜单。)我想点牛排、薯条、蔬菜沙拉。

A：请问您的牛排是要三分熟、五分熟还是全熟的?

B：五分熟吧! 我能看看酒单吗?

A：好的。

B：来瓶葡萄酒。

A：女士，还需要别的吗?

B：果汁。就这些吧!

A：好的，请稍等。

B：谢谢!

How to Write Expository Essay

Definition

Expository essay is an article that explains and clarifies things in the way of expository expression, with the purpose of explaining and describing the shapes, characteristics, properties, causes, relations, functions and effects of things. Explanation is a basic expression of the occurrence, development, results, characteristics, properties, states and functions of things.

Classification

There are many writing techniques that can be used to explain things. Four major types of expository texts are introduced as follows:

(1) Illustrate a point by examples in order to make the readers understand a certain method.

(2) Introduce facts, information, knowledge, etc. in order to make the readers gain insight and know what to do.

(3) Describe the similarities or differences between two things by comparing them in order to make the readers learn from the comparison.

(4) Reveal the nature and characteristics of things by defining in order to make the readers understand their exact meanings.

In practical writing, the author usually uses multiple methods to achieve the writing purpose of explaining things.

Whatever methods are used, the aim is to keep the expository essays clear, appropriate, accurate, and concise.

【Sample】

How to Plant Sweet Melon

1. Seed selection: select healthy plump particles.

2. Seed soaking: put the seeds under the sun for disinfection, and soak them in warm water for 5 hours.

3. Seed germination: dry the seeds and mix wet sand in a seedling bowl for moisturizing.

4. Ditch digging: dig the plot deeply and dig the planting ditch.

5. Fertilizer application: put decomposed human and animal feces and urine as the base fertilizer into the planting ditch.

6. Seed sowing: sow the seeds into the ditch.

7. Soil covering: cover the planting ditch with fine soil of 2-3 centimeter height.

8. Soil watering: wet the soil with sufficient water to keep it moist.

9. Seed transplanting: when three healthy leaves grow out of a melon seed, transplant them with soil into a loose plot with a sunny leeward; May is the best transplanting time; 15-centimeter apart is the best row spacing of plants.

10. Post-management: awnings should be set up to avoid strong sunlight; the soil should be loosened to prevent compaction; weeds should be removed in time for the growth of seedlings.

✔ Assignment: *Write a simulated expository essay on how to use agricultural harvesters.*

Unit Two
Artificial Intelligence in the Agriculture Industry

Study for Fun

Enjoy English and Relax

Directions: *Can you guess the metaphoric meanings of the human body parts?*

brain	脑袋	You have a lot of brains!
back	后背	see one's back
eye	眼睛	Mind your eye!
ear	耳朵	Walls have ears!
nose	鼻子	nose to nose
tongue	舌头	mother tongue
face	脸颊	about face
chin	下巴	Keep your chin up!
shoulder	肩膀	turn a cold shoulder
body	身躯	Little bodies may have great souls!
chest	心胸	community chest
kidney	肾脏	man of the right kidney
lap	膝盖	in the lap of fortune
lung	肺部	You have good lungs!
breast	胸膛	a troubled breast

Unit Three

Agriculture Outlook

Lead-in

Discussion: What is the Difference Between Organic Agriculture and Traditional Agriculture?

● Organic agriculture is an agricultural method that maintains the development of agricultural production mainly by organic fertilizer, crop rotation and mechanical tillage, while avoiding the use of pesticides and chemical fertilizers in agricultural production.

● Traditional agricultural production systems mainly rely on the input of pesticides and fertilizers to control pests, diseases, and weeds. In organic agriculture mode, livestock manure, green manure, leguminous crops and organic wastes are used to maintain soil productivity, crop rotation, timely sowing and cultivation. Biological measures are also used to control pests and diseases.

Intensive Reading

Advances and Prospects of Super Rice Breeding in China

Super rice breeding in China has been very successful over the past three decades, and the Chinese government has made progress in the potential of super rice breeding. After the establishment of the breeding theory and strategy of "generating a kind of rice with strong **heterosis** through inter-**subspecies hybridization**, by examining gene to combine elite traits through **composite**-crossing to breed super rice varieties with both ideotype and strong hybrid vigor", a series of major **breakthroughs** have been achieved in both ordinary and super hybrid rice breeding. A number of new genetic materials with ideotype have been created successfully. With widespread **cultivation** of super rice with higher quality and yield, as well as **resistance** or tolerance to **abiotic** or **biotic** stresses, the yield of rice per unit has reached a new level. In addition to increased quality and yield, hybrid rice breeding has also led to improvements in many other traits, such as resistance to pests and diseases, resistance to lodging and efficient light **distribution** in population. Achievements in super rice breeding and **innovation** in rice production have made major **contributions** to the progress in rice sciences and worldwide food security.

Food security is a key issue concerning China's national economy and people's livelihood. Maintaining grain self-sufficiency in China has become harder as the population grows and farming land **shrinks**, so the pressure on food security is increasingly severe. Rice is a **dietary** food for more than half of China's population, so improving rice production is crucial for ensuring food security in China. Historical experience has shown that development of gene resources in rice breeding is an effective way to increase the production potential of certain rice varieties. These genetic resources can then be used, in combination with newly developed technologies, to improve rice yields.

In the past, breeding of **dwarf** varieties and hybrid rice led to increased yields. In the 1980s, super rice, a **concept** of varieties with very high yield, was proposed as a new breeding purpose. Over thirty years of breeding research in super rice has resulted in successful development of new **germplasm** resources, new varieties, and has improved cultivation techniques. These significant breakthroughs have made contributions to both national food security in China and progress of rice science.

Super rice breeding started in the rice cultivation area of Northeast China. Through years of work, the northeastern super rice cultivation areas developed from small to large **scale**, turning the theory of yield improvement into reality with increased rice yield. During this period, super rice breeding was spread nationwide. In 2011, the growing areas of super rice varieties reached about 24.7% of the total rice cultivation areas in China. Super rice varieties provided 13.2% more production than common rice varieties and contributed an **estimated** increase of 7.5 billion kg in rice production in China.

So far, **remarkable** achievements has been made in super rice breeding. During the 13th Five-Year Plan Period, rice production continued to grow. However, with the economic development in China in the past four decades, the cost of arable land and labor has increased, and thus the price of domestic food production has risen, facing the pressure of international low prices. In this context, super rice, which **enhances** productivity per unit area, has contributed to the improvement of food production in China. Consequently, the strategy of super rice breeding in China should pay equal attention to increasing yield, improving quality and reducing costs under the **premise** of safe production. Super rice breeding in China is expected to make greater contributions to China's food security in the near future.

Word Bank

heterosis [ˌhetəˈrəusis]	n.	the increased size, strength, etc. of a hybrid as compared to either of its parents 杂交优势
subspecies [ˈsʌbspiːʃiːz]	n.	one species derived from another（动植物）亚种
hybridization [ˌhaibridaiˈzeiʃn]	n.	bred from two species 杂交种；混合物
composite [ˈkɔmpəzit]	a.	mixed together 合成的；混合的
breakthrough [ˈbreikθruː]	n.	an important achievement 突破；进展
cultivation [ˌkʌltiˈveiʃn]	n.	production of food or crops 耕种；栽培

Unit Three
Agriculture Outlook

resistance [ri'zistəns]	n.	refusal to obey 抵制；抵抗
abiotic [,eibai'ɔtik]	a.	not involving biology 与生物无关的；非生物的
biotic [bai'ɔtik]	a.	relating to living organisms 生物的；生命的
distribution [distri'bju:ʃn]	n.	the act of giving sth. to different people or places 分发；分布
innovation [,inə'veiʃn]	n.	a new thing or a new method 创新；新方法
contribution [,kɔntri'bju:ʃn]	n.	the act of giving help to produce 贡献；奉献
shrink [ʃriŋk]	v.	become smaller in size 缩水；使缩小
dietary ['daiətəri]	a.	of food that concerns a person's diet 饮食的
	n.	a regulated daily food allowance 规定的食物；食谱
dwarf [dwɔ:f]	v.	to make sth. seem very small 使……显得过于矮小
	a.	(of a plant or an animal) smaller than usual size（植物或动物）矮小的
concept ['kɔnsept]	n.	an idea or abstract principle 概念；观念
germplasm ['dʒə:mplæzm]	n.	a collection of genetic resources for an organism 胚质；生殖质
scale ['skeil]	n.	size or proportion 规模；比例
estimate ['estimeit]	v.	judge or calculate 判断；估计
remarkable [ri'ma:kəbl]	a.	unusual or special 不同寻常的；引人注目的
enhance [in'ha:ns]	v.	improve to a higher level 提高；增强
premise ['premis]	n.	a statement assumed to be true 前提；假定

Extended Reading

Table Etiquettes

Table manners are the way you behave when having a meal. To have good table manners, there are some rules to follow. Before eating, please put a napkin on the lap, but never tuck into any part of the clothes. You should not begin to eat until your host or hostess has begun. Sit up straight on your chair. Chew slowly with your mouth closed. As long as there is food in the mouth, do not try to talk. If you have to cough, use your napkin to cover your mouth.

Things one should avoid doing:

✗ Do not clean your teeth at the table with a toothpick, or your finger, or even your tongue.

✗ Do not lean against the back of the chair or lean on the table with your elbows.

✗ Do not put too much food in your mouth at a time.

✗ Do not make any noise with your mouth while eating.

Avoid talking about bad topics at table:

✗ Waste passed from the body.

✗ Anything connected with the private parts of the body.

✗ Sex relations.

✗ Age, income or salary, marital status and religious beliefs.

Avoid rude behavior at table:

Spitting on the ground, yawning, sniffling, passing gas, belching out loud, doing ear cleaning, picking one's nose, scratching an itch, talking loudly without paying attention to others, smoking without permission, laughing at someone who makes mistakes in public, having an extra long nail on the little finger, going around examining things without permission, not remaining standing when a lady comes nearby the table, etc.

Unit Three
Agriculture Outlook

Exercises

I. Reading Comprehension

Directions: *Choose the best answer for the following questions.*

1. Over the past three decades, a series of major breakthroughs have been achieved in _____ . (   )

    A. many other agronomic traits

    B. the rice production technology

    C. the yield of rice production per unit

    D. conventional and super hybrid rice breeding

2. Which factor helps the rice sciences and worldwide food security? (   )

    A. The new method in rice production.

    B. Progress in rice breeding.

    C. Increasing in rice quality and yield.

    D. Improvements in many other agronomic traits.

3. Which of the following statements is true? (   )

    A. Numerous methods can be used to increase rice production.

    B. The pressure on food security has little relation with arable land shrinks in China.

    C. There will be two ways to increase the grain output of rice.

    D. Rice is the principal food for more than half of China's population.

4. Where did super rice breeding start? (   )

    A. In the rice cultivation area of Southeast China.

    B. In the rice cultivation area of Northeast China.

    C. In the rice cultivation area of Northwest China.

    D. In the rice cultivation area of Southwest China.

5. Which circumstance is not related to the contribution of super rice in China in the past years? (   )

    A. Economic development in China in the past four decades.

    B. The increased cost of arable land and labor.

    C. More expensive domestic food production.

D. The pressure from low domestic prices.

Ⅱ. Vocabulary

Section A

Directions: *Match the English words or phrases with their Chinese equivalents.*

1. (　) super rice breeding	A. 第十三个五年规划
2. (　) gene pyramiding	B. 水稻单产
3. (　) new genetic materials	C. 在……方面做出贡献
4. (　) higher quality and yield	D. 粮食安全
5. (　) rice production per unit	E. 基因金字塔法
6. (　) food security	F. 优质高产
7. (　) genetic resources	G. 遗传资源
8. (　) make contributions to	H. 新型遗传材料
9. (　) make achievement	I. 取得成就
10. (　) the 13th Five-Year Plan	J. 超级稻育种

Section B

Directions: *Translate the following words and phrases.*

1. 压力
2. 超级
3. 粮食产量
4. 提升
5. 同时地
6. 品种
7. 推广
8. 幼苗多样性
9. 基因
10. 降低
11. varieties
12. simultaneously
13. grain output
14. super
15. spread
16. pressure
17. enhance
18. dwarf
19. gene
20. automatic grafting machine

Ⅲ. Situational Speaking
Doing Overseas Business

A: Hello! It's nice to meet you. I'm Li Ming from Beijing.

B: It's a pleasure to meet you. Here's my card. I'm Wang Jun.

A: Director of marketing?

B: Yes.

A: OK! And here's my card.

B: Would you like something to drink?

A: Thank you! A cup of tea, please.

B: So shall we go over the agenda?

A: OK! And your offer directions were very straightforward.

B: Any suggestions?

A: Our company wants to know your final offer price.

B: We'll have them worked out by the evening and let you have them tomorrow morning.

A: If you can reduce the price by 5%, we shall order 100 metric tons.

B: Business is possible if you can increase the price by 2%.

A: Our offer price is net without commission.

B: Our price is highly competitive when you consider the quality.

A: If your price is agreeable, and we can place the order right now.

B: I'm sure you'll find our price most favorable.

A：你好！见到你很高兴！我是来自北京的李明。

B：见到你也很高兴。这是我的名片。我是王军。

A：市场营销总监？

B：是的。

A：好。这是我的名片。

B：你想喝点什么吗？

A：谢谢！请来杯茶。

B：我们讨论一下议题吧？

A：行。你们的报价方向已经很明确了。

B：您有什么建议吗？
A：我们公司想知道最终的报价。
B：今晚就会最终出炉，最迟明天上午给贵公司。
A：如果价格可以降低5%，我们会订购100吨。
B：如果贵公司提价2%，交易才有可能。
A：我们的价格是净价，不含佣金服务费。
B：如果贵公司考虑质量的话，我们的价格是很有竞争力的。
A：如果价格合适，我们可以马上签单。
B：你们会发现我们公司的价格是最优惠的。

Writing

How to Write Expository Notice and Announcement

Notices and announcements are written or orally stated, making others know what has happened or what will happen, and when it happened or when it will happen. Besides, to whom the information is given should be included. The language of notices and announcements should be concise, simple, accurate and somewhat formal.

Notices can be written on blackboards or bulletin boards. They can also be written as memos delivered to the officials or as letters to notify people of something in detail.

An announcement is a public or official statement that gives people information about something. In general, announcements can be delivered in written or oral forms. Written or printed announcements are usually published in newspapers, or posted at airports, in stations, etc.

【Sample 1】

Announcement

The swimming pool of Xinhua University will be open to the public from June 15th. Please bring your own swimsuits.

Time: 8:00 a.m.-10:00 p.m.

Unit Three
Agriculture Outlook

Fee: RMB 3 Yuan / hour for an adult
RMB 2 Yuan / hour for a child

【Sample 2】

Notice

All professors and associate professors are requested to meet in the college conference room at 2:00 p. m. on Thursday (August 18th) to discuss issues of international academic exchanges.

The Headmaster's Office
August 15th, 2022

Assignment: *Write a simulated notice of International Labor Day Holiday of your farm.*

_____
_____
_____
_____
_____
_____

Study for Fun

Relax and Enjoy the Song

Directions: *Some words are missing in the song. While you are listening, please fill in the missing words.*

Yesterday Once More

When I was young I'd listen to the radio
Waiting for my favorite _____
When they played I'd sing along
It made me _____
Those were such happy times and not so long ago
how I wondered where they'd gone.

But they're back again just like a long lost friend
All the songs I _____ so well
Every sha-la-la-la, every wo-o-wo-o
Still shines
Every shing-a-ling-a-ling
That they're starting to _____
So fine
When they get to the part
Where he's breaking her _____
It can really make me cry
Just like before
It's yesterday once more
Every sha-la-la-la, every wo-o-wo-o
Still shines
Every shing-a-ling-a-ling
That they're starting to sing
So fine
When they get to the part
Where he's breaking her heart
It can really make me cry
Just like before
It's yesterday once more

Unit Four

Ecological Agriculture

Lead-in

Discussion: What are the Obstacles to the Agricultural Resource Conservation?

● The destruction of agricultural environment caused by some enterprises and farmers in pursuit of higher economic profits.
● Insufficient awareness of agricultural resource conservation.
● Lack of agricultural resource conservation professionals.
● Inadequate publicity of agricultural resource conservation.
● Insufficient awareness of environment protection.
● Inadequate implement of institutional policies.
● Inadequate investment of agricultural resource conservation facilities and research.
● Imbalance between planting effect and expanding speed of crop planting areas.
● Insufficient professional technicians.
● Severe pollution.

Intensive Reading

Antioxidant Effect of Glutathione in Vitamin C Fermentation System

Vitamin C is an essential water-soluble vitamin for mammals, and has been widely used for a long time in **pharmaceutical**, food and **cosmetic** industries because of its **antioxidant** property. Two-step microbial **fermentation** method has been widely used in Vitamin C industrial production. Many researches on Vitamin C production have focused on the construction of engineered strains. For the companion strains, the formation and release of cells significantly promoted the growth and enhanced the production.

Lysozyme was used specifically to release **intracellular metabolites** into Vitamin C fermentation system, which improved the growth of industrial production. The studies on the release of intracellular **substances** of companion strain are based on experiments. However, the effects of adding **glutathione** into Vitamin C fermentation system have not been reported.

The activities of antioxidant-related **enzymes** were used to study the antioxidant effect of glutathione addition to Vitamin C fermentation system. The addition of glutathione increased 2-KGA（一种糖酸）production and decreased fermentation time. The highest 2-KGA production was increased by 40%-63%, and the lowest fermentation time was shortened to 60 hours when the addition of **optimal concentration ratio** of glutathione was 50∶1. Moreover, the increased production of 2-KGA was **accompanied** by up-regulated activities of total antioxidant **capacity**（T-AOC）, total **superoxide dismutase**（T-SOD）and over-expressed **oxidative** stress-related genes, which resulted in **wiping** out reactive oxygen **species** to reduce oxidative stress in Vitamin C fermentation system.

Experimental results suggested that the addition of glutathione could **significantly** change the oxidation-reduction state and provided a favorable reduced state for 2-KGA production in Vitamin C fermentation system.

Data **indicated** that the addition of glutathione could **respond** to oxidative stress and improve the oxidation-reduction state of Vitamin C fermentation system.

Unit Four
Ecological Agriculture

In order to further **explore** the **mechanism** of glutathione to enhance 2-KGA production and **synergistic** effects between glutathione, the related enzymatic **dynamic** activities of intracellular **fluid** and extracellular fluid in Vitamin C fermentation system were also **measured**. To further **investigate** the effect of glutathione on Vitamin C fermentation system, the expression of oxidative stress-related genes was **analyzed**.

It is concluded that glutathione showed a significant effect on **decreasing** fermentation time in Vitamin C fermentation system. Its significance and impact is noticeable. Glutathione is proved to be effective to relieve oxidative stress. Its promotion effects help to further explore the fermentation processes of industrial production.

Word Bank

pharmaceutical [fɑːməˈsjuːtikəl]	a.	connected with making medicines 制药（学）的
cosmetic [kɔzˈmetik]	n.	a substance for make-up 化妆品
antioxidant [æntiˈɔksidənt]	n.	substance that inhibits oxidation 抗氧化剂
fermentation [ˌfəːmenˈteiʃn]	n.	a chemical change with effervescence 发酵；发酵变化
lysozyme [ˈlaisəzaim]	n.	[医] 溶解酵素
intracellular [intrəˈseljulə]	a.	situated inside a cell 细胞内的
metabolite [meˈtæblait]	n.	any substance involved in metabolism [生化] 代谢物；代谢分子
substance [ˈsʌbstəns]	n.	a type of solid, liquid or gas that has particular qualities 物质；物
glutathione [gluːtəˈθaiəun]	n.	[生化] 谷胱甘肽；促性腺激素
enzyme [ˈenzaim]	n.	[生化] 酶
optimal [ˈɔptiməl]	a.	the most desirable 最佳的；最优的
concentration [kɔnsnˈtreiʃn]	n.	the amount of a substance in a liquid 浓度

ratio [ˈreiʃiəu]	n.	the relation between two groups 比率；比例
accompany [əˈkʌmpəni]	v.	go somewhere with sb. 陪伴；伴随
capacity [kəˈpæsəti]	n.	the amount of containing 容量
superoxide [ˌsjuːpəˈɔksaid]	n.	[化学] 超氧化物；过氧化物
dismutase [disˈmjuːteis]	n.	[化学] 歧化酶
oxidative [ˈɔksideitiv]	a.	easily rusty [化学] 氧化的
wipe [waip]	v.	remove or rub from a surface 擦；抹
species [ˈspiːʃiːz]	n.	a group of animals, plants [生物] 物种；种类
experimental [ikˌsperiˈməntl]	a.	based on new ideas, forms or methods that are used to find out what effect they have 以实验（或试验）为基础的；实验性的；试验性的
significantly [sigˈnifikəntli]	ad.	in an important noticed way 显著地；明显地
indicate [ˈindikeit]	v.	show that sth. is true or exists 表明；预示
respond [riˈspɔnd]	v.	do sth. as a reaction to sth. 反应；回应
explore [ikˈsplɔː]	v.	examine in order to find out sth. 探索；探究
mechanism [ˌmekənizəm]	n.	a method or a system for achieving sth. 方法；机制
synergistic [ˌsinəˈdʒistik]	a.	cooperative or working with sb. 协同的；协作的
dynamic [daiˈnæmik]	a.	energetic and active 有活力的；精力充沛的
fluid [ˈfluːid]	n.	a liquid or substance that can flow 流体；液体
	a.	flowing freely as liquids do 流动的；液体的

measure ['meʒə]	v.	find the size or quantity of sth. 测量，判断
investigate [in'vestigeit]	v.	examine or find out carefully 调查；研究
analyze ['ænəlaiz]	v.	examine the nature of sth. 分析；研究
decrease [di'kri:s]	v.	become smaller in number 减少；缩减
	n.	the amount of sth. reduced 减少量；缩减量

Extended Reading

Hand-Shaking Etiquettes

It is believed that handshaking originated from the habit of ancient warriors. People usually use weapons with their right hands. Extending their unarmed right hands to their rivals shows that they have no intention of making an enemy. It is a symbol of friendship.

It is customary to shake hands when people meet each other. People also shake hands when parting in business and other formal situations. Friends also shake hands when they meet after having not seen each other for a long time. If people meet quite often, they may make a little bow to each other instead of shaking hands. If others offer to shake hands, it would be very bad manner to refuse.

Proper hand-shaking etiquettes are as follows:

• Between people of the same sex, it is usual for the older to put his or her hand out first.

• Between different sexes, it is the woman who first offers her hand. If she prefers not to shake hands, she bows slightly.

• If a man is wearing gloves, he should take off the glove from his right hand before shaking hands. If there is any difficulty of hand-shaking, he must say, "Excuse my glove." A woman does not take off her glove.

• When a man shakes hands with a woman, it is only a loose shake.

• When a man shakes hands with another man, they should shake hands firmly.

• Don't shake hands with your left hand.

• Don't shake hands with the opposite sex with both hands.

Exercises

I. Reading Comprehension

Directions: *Choose the best answer for the following questions.*

1. What is the solution suggested by the author to explore antioxidant effect of glutathione in Vitamin C fermentation system? (　　)

　　A. The promotion effects of glutathione on 2-KGA production could be helpful.

　　B. Policy makers should aim at establishing and improving mechanisms.

　　C. The structural reform of the biological supply-side should be accelerated, and well-coordinated production should be pursued.

　　D. The two-step microbial fermentation method should be understood.

2. Which of the following statements can be identified as a defect of Vitamin C fermentation system? (　　)

　　A. Glutathione showed a significant effect on increasing 2-KGA production and decreasing fermentation time in Vitamin C fermentation system.

　　B. The effects of adding glutathione into Vitamin C fermentation system have not been reported.

　　C. Many researches on Vitamin C production have focused on the construction of engineered strains.

　　D. The relevant pathways deficiencies are based on experiments.

3. What attitude of the author can be inferred towards Vitamin C fermentation system research? (　　)

　　A. Indifferent.

　　B. Timely.

C. Active.

D. Ignorant.

4. Which of the following statements is true in order to investigate antioxidant effect of glutathione in Vitamin C fermentation system? (   )

    A. The activities of antioxidant-related enzymes were used to study the antioxidant effect of glutathione addition on Vitamin C fermentation system.

    B. The studies on the release of intracellular substances of companion strain and the relevant pathways deficiencies were conducted.

    C. Engineered strains were constructed to study the antioxidant effect of glutathione addition on Vitamin C fermentation system.

    D. Lysozyme was used specifically to damage cell wall structure and release intracellular metabolites into Vitamin C fermentation system.

5. What does the word "dynamic" mean in the sentence "... the related enzymatic *dynamic* activities of intracellular fluid and extracellular fluid in Vitamin C fermentation system were also measured"? (   )

    A. Inactive.

    B. Efficient.

    C. Static.

    D. Forceful.

Ⅱ. Vocabulary

Section A

Directions: *Match the English words or phrases with their Chinese equivalents.*

1. ( ) microbial fermentation	A. 缩短发酵时间
2. ( ) antioxidant property	B. 协同作用
3. ( ) enhance production	C. 保持动态平衡
4. ( ) shorten fermentation time	D. 微生物发酵

5. (　) explore the mechanism	E. 对……有显著作用
6. (　) synergistic effects	F. 抗氧化性
7. (　) oxidative stress-related genes	G. 过程优化
8. (　) to have a significant effect on sth.	H. 研究机制原理
9. (　) maintain a dynamic balance	I. 提高产量
10. (　) the optimization of process	J. 氧化应激相关基因

Section B

Directions: *Translate the following words and phrases.*

1. 耐抗生素
2. 基因编辑
3. 生物技术
4. 人工诱变
5. 无土栽培
6. 经济作物
7. 智能温室
8. 高端种植
9. 生态农业
10. 农业资源保护

11. humid acid
12. meadow
13. biotechnology
14. chromosome
15. soilless cultivation
16. amine salt
17. arable land
18. mechanized farming
19. pasture land
20. agricultural resource conservation

Ⅲ. Situational Speaking

At Bank

A: What can I do for you?

B: Please tell me my balance.

A: Can I see your ID, your card and the withdrawal slip, please?

B: Sure.

A: Your balance at the card is 5,000 yuan.

B: I would like to withdraw 2,000 yuan from my account.

A: Let me just make sure. You want to withdraw 2,000 yuan.

Unit Four
Ecological Agriculture

B: That's right.

A: Can I help you with other things?

B: I need to cancel my accounts.

A: Is there a problem with it?

B: I don't need it anymore.

A: A deposit or current account?

B: Both.

A: What would you like to do with all the money in this account?

B: Just transfer it over to my remaining card.

A: It's going to take a moment for me to cancel your account.

B: That's fine. Take your time.

A: Thank you!

A：我能为您做些什么？

B：请告诉我结余金额。

A：我能看一下您的身份证、银行卡和取款单吗？

B：当然。

A：您卡上的余额是5 000元。

B：我想从账户中支取2 000元。

A：我来确认一下。您要取2 000元。

B：没错。

A：我还能帮您做些别的吗？

B：我需要注销我的账户。

A：这个账户有问题吗？

B：我不再需要它了。

A：定期还是活期账户呢？

B：两者都注销。

A：您想如何处理这个账户里的钱？

B：转到我剩余的卡上就行了。

A：我要花点时间注销您的账户。

B：好的。不着急。

A：谢谢您！

Writing

How to Write Business Letters

There are three basic formats of laying out business letters: block (齐头式), modified block (混合式) and indented format (缩进式). Business letters often include 10 elements.

1. Letterhead（信头）

The letterhead stays on top of the business letter, showing the company's name, full address, postal code, telephone number, fax number, email and website.

2. Date（日期）

The date is usually placed at least one blank line below the letterhead. It can be marked by "Day Month Year" (e.g., 20th August 2020) or "Month Day, Year" (e.g., August 20th, 2020).

3. Inside Address（封内地址）

The inside address is one or more lines below the date. It shows the name and the address of the receiver. "Mr." or "Ms." is usually used (e.g., Mr. Ming Lin). If it is not an official letter, the inside address may be omitted.

4. Salutation（称呼）

The salutation, also called greeting, shows the sender's respect for the receiver. "To Whom It May Concerns" is a fashionable way of salutation.

5. Opening（开头）

The opening should be relevant to the subject or to the reader. In other words, it is better for the readers to learn the purpose of the letter as early as possible.

6. Body（正文）

The body of the letter deals with the reasons for writing the letter in detail. Paragraphing is necessary. Try to be concise, direct and clear.

7. Closing（结尾）

The closing paragraph performs two functions: summarizing the main points and requesting the action to be taken if necessary.

8. Complimentary Close（信尾客套语）

It is placed on the second line (double space) below the closing of the letter. Commonly used are: "Sincerely yours", "Faithfully yours", "Sincerely", etc.

9. Signature（签名）

The sender's signature is written below the complimentary close.

10. Enclosures（附件）**and Carbon Copy**（抄送）

If any document is attached to the business letter, type "Enc." at the bottom.

[Sample]

Ning Tian
No. 24 Main Street
Linxia, Gansu Province
(0086) -0930-1246××××
August 20th, 2020

Dear Mr. Lin,

After seeing your advertisement of Oriental S6 Series tractors at the 10th Regional Modern Agricultural Exhibition, we are impressed with the good quality of the tractors.

We are an export company of agricultural machinery. Recently, one of our clients is in need of mini-type tractors, so we shall be delighted to establish business relations with you. Could you please send us your catalogues and a price list concerning the above-mentioned tractor products?

For more information concerning our credit standing, please refer to Linxia Branch of Agricultural Bank of China.

We look forward to your reply.

Sincerely,
Ning Tian
CEO, Linxia Export Company of Agricultural Machinery

Enc.: Company Introduction

Assignment: *Write a simulated business letter to establish business relations with your potential partner.*

Study for Fun

Enjoy English and Relax

Directions: *Match the proverbs and idioms below.*

轻如鸿毛	close as an oyster
坚如磐石	cool as a cucumber
噤若寒蝉	firm as a rock
守口如瓶	as strong as a horse
泰然自若	mute as a fish
强壮如牛	light as a feather
笑里藏刀	safe and sound
针锋相对	tit for tat
不伦不类	neither fish nor fowl
安然无恙	A fair face may hide a foul heart!

Unit Five

Agricultural Product Processing

Lead-in

Discussion: What do You Know about Agricultural Processing?

● Agricultural processing spans three major fields of agriculture, industry and service industry. It has the characteristics of less investment, short cycle and quick return, and is a priority industry for developing countries. The practice of developed countries and economically developed areas in China shows that agricultural processing industry can extend the agricultural industrial chain, increase the added value of agricultural products, increase farmers' income and reduce the post-production losses of agricultural products.

● There are still some problems in agricultural processing. For example, agricultural processing industry develops rapidly on the whole, but its position in the industry declines; the export value showed an upward trend, but its share declined; the technological level needs to be improved; the variety of processing machinery can only meet the initial demand; the tax policy of agricultural processing industry is not optimized enough; the financial input on mechanism is not adequate; the processing equipment is not advanced enough; there is a shortage of professional and technical personnel; the scale of processing enterprises is not large enough.

● Measures should be taken to enhance agricultural efficiency, develop rural prosperity and increase the income of farmers.

Intensive Reading

Tomato Processing and Pesticide Residues

In recent decades, tomato is one of the most widely grown vegetable plants in China and is also a popular raw material of Chinese dishes. Tomatoes are consumed both as fresh vegetables and as processed products **including** tomato juice and tomato **paste**.

Tomatoes are **susceptible** to diseases in the growing period, for instance, **phytophthora infestans**. Cyazofamid is a relatively novel **fungicide** which has specific activity against diseases caused by **oomycetes** and can inhibit a broad spectrum of plasmodiophoromycetes and oomycetes. Cyazofamid has been widely used to control diseases in fruits and vegetables, such as tomato late blight. Regueiro's study points out that cyazofamid was **toxic** to **cortical neuron** cells, resulting in a significant reduction in cell survival, and that it could also induce **depolarisation** of mitochondrial **membrane**. Subchronic toxicity test showed that cyazofamid had harmful effects on the kidneys of male rats. Cyazofamid rapidly **decomposes** into CCIM after field application. Since CCIM is more easily absorbed and more toxic than cyazofamid, its **residues** in agricultural products may lead to higher risk from dietary intake. Meanwhile, a large number of cyazofamid applications associated with vegetable production may raise the potential risk for human exposure and make a harmful impact on the environment. Therefore, **monitoring** the fate of cyazofamid and CCIM in tomato is necessary for human health and environmental protection.

Home-canned tomato pastes are indispensable processed food for many families, and the level of pesticide residues in food is uncertain during home processing. In recent years, many studies have shown the influence of processing on pesticide residues in agricultural products. Usually food processing techniques (e. g. washing, **peeling**) can lead to some decrease of pesticide residues in products. Though some techniques including drying or concentration may result in an increase of residues in products, it is important

Unit Five
Agricultural Product Processing

to **evaluate** the **exposure** level of pesticide residues in different processed foods. Moreover, processing factor is a main **parameter** to assist dietary intake of pesticides in processed agriculture products. To our knowledge, many papers have reported the residues of cyazofamid and its metabolite, CCIM, in the fields, but little attention has been given to the levels of their residues caused by home or commercial preparation. Thus, in order to ensure food safety for **consumers**, we carried out a field experiment on tomato and tomato paste processing.

Washing is an important step for reducing pesticide residues. Research has found that the unwashed tomato skin had the largest amount of concentrations of cyazofamid and CCIM. This is probably because the cyazofamid SC first **attaches** on the tomato surface after **spraying**.

By comparing the residues of cyazofamid in unwashed tomato skin and peeled tomatoes, it was found that the residues of cyazofamid in the former were much higher than those in the latter, which indicated that **cutin** and **wax** may play an important role in physically protecting tomato fruit from pesticide deposition. In addition, the research result also fully proved that peeling was more effective than washing in removing pesticide residues of cyazofamid because the peeling step not only removes pesticide residues on the surface of tomatoes but also **eliminates** pesticides which have **penetrated** into the skin of tomatoes.

Word Bank

include [inˈkluːd]	v.	consider as part of sth. 包含；把……列为一部分
paste [peist]	n.	a smooth mixture of crushed meat, fish, etc. 肉（或鱼等）酱（作涂抹料或烹饪用）
susceptible [səˈseptəbl]	a.	very likely to be influenced 易受影响的
phytophthora [faiˈtɔfθərə]	n.	destructive parasitic fungi in plants 疫霉

fungicide ['fʌngisaid]	n.	any agent that destroys or prevents the growth of fungi [药]杀真菌剂
oomycetes [ˌəuəˈmaisi:ts]	n.	nonphotosynthetic fungi that resemble algae [微] 卵菌（一类真菌）
toxic ['tɔksik]	a.	poisonous 有毒的；引起中毒的
cortical ['kɔ:tikl]	a.	of or relating to a cortex 皮质的
neuron ['njuərɔn]	n.	a cell that carries information within the brain 神经元
depolarisation ['di:pəulərai'zeiʃn]	n.	loss of polarity or polarization 去极化（作用）
membrane ['membrein]	n.	a very thin layer found in the structure of cells in plants（植物的）细胞膜
decompose [ˌdi:kəm'pəuz]	v.	separate (substances) into constituent parts 分解；降解
residue ['rezidju:]	n.	matter that remains 剩余物，残留物
monitor ['mɔnitə(r)]	v.	check, track or observe sth. 监管；管理
peeling ['pi:liŋ]	n.	loss of outer skin by shedding 剥落；脱落
evaluate [i'væljueit]	v.	form an opinion of sth. 评价；评估
exposure [ik'spəuʒə(r)]	n.	the state of being exposed 暴露；接触
parameter [pə'ræmitə(r)]	n.	sth. that limits the range 规范；范围
consumer [kən'sju:mə(r)]	n.	a person who buys goods or uses services 消费者；购买者
attach [ə'tætʃ]	v.	fasten one thing to another 固定；依附
spray [sprei]	v.	cover sth. with drops of a liquid 喷洒
cutin ['kju:tin]	n.	a waxy waterproof substance 角质
wax [wæks]	n.	a solid substance that is made from fats and oils and used for making candles, polish, models, etc. 蜡

| eliminate [iˈlimineit] | v. | remove or get rid of sth. 排除；清除 |
| penetrate [ˈpenitreit] | v. | go into or through sth. 穿过；进入 |

Extended Reading

Business Dressing Etiquettes

Suits

When choosing a suit, firstly, consider material, fitness, and comfort. Secondly, consider style. The material should never shine or change colors in different lighting. Appropriate materials include wool, cotton, and linen.

Shoes

Coordinate your shoe color with your suit color. Black shoes go nicely with charcoal and black or navy suits. Brown shoes match with brown and tan suits.

Ties

• Ties should be silk.
• Ties should be understated.
• The color of the tie should be coordinated with that of the suit and shirt.
• The end of the tie should be above your belt.

Business casual and Friday casual are distinct dressing. Business casual generally includes khaki pants, a plain polo shirt or a long-sleeved button shirt. Friday casual generally includes a V-neck sweater, sometimes a sports coat or jacket, and brown leather shoes.

Exercises

Ⅰ. Reading Comprehension

Directions: *Choose the best answer for the following questions.*

1. Which of the following statements is right according to the text? (   )

A. The tomato is a kind of ornamental plant.

B. Tomatoes are rarely grown in China.

C. Tomatoes are consumed both as fresh vegetables and as processed products.

D. Tomatoes cannot be processed.

2. Tomatoes are susceptible to diseases in the growing period, for instance, _____ . (   )

A. aphid

B. fungous disease

C. powdery mildew

D. phytophthora infestans

3. Food processing techniques including _____ may result in an increase of residues in products. (   )

A. eating

B. drying or concentration

C. washing

D. peeling

4. Besides washing, which of the following is an important step to reduce pesticide residues? (   )

A. Heating.

B. Drying.

C. Peeling.

D. Frying.

5. Cutin and _____ may play an important role in physically protecting tomato fruit from pesticide deposition. (   )

A. water

B. wax

C. air

D. sunlight

Ⅱ. Vocabulary

Directions: *Match the English words or phrases with their Chinese*

Unit Five
Agricultural Product Processing

equivalents.

1. ( ) raw material		A. 加工产品
2. ( ) processed products		B. 蔬菜生产
3. ( ) fresh vegetable		C. 番茄酱
4. ( ) vegetable production		D. 农产品
5. ( ) pesticide residues		E. 原料
6. ( ) tomato paste		F. 食品安全
7. ( ) tomato juice		G. 潜在风险
8. ( ) agriculture products		H. 番茄汁
9. ( ) food safety		I. 新鲜蔬菜
10. ( ) potential risk		J. 农药残留

Section B

Directions: *Translate the following words and phrases.*

1. 农药
2. 番茄酱
3. 田间试验
4. 神经细胞
5. 农产品
6. 食品加工
7. 蔬菜表皮
8. 清洗
9. 去皮
10. 烘干
11. infestan
12. environmental protection
13. vegetable production
14. fungicide
15. consumer
16. decompose
17. dietary
18. evaluate
19. spray
20. penetrate

III. Situational Speaking
Likes and Dislikes

Section A

A: Hello, Fangfang.
B: Hello, Xiaoming.
A: Hey, what is your favorite food?

B: My favorite food? Well, in China it's probably dumplings.

A: How often do you have dumplings?

B: Probably at least two times a week.

A: What is your favorite drink?

B: My favorite drink? Well, tea usually.

A: Do you have tea every day?

B: No, because I will lose sleep if I drink too much.

A：你好，芳芳。

B：你好，小明。

A：你最喜欢什么食物啊？

B：我最喜欢的食物？嗯，在中国的话，大概是饺子。

A：那你多久吃一次饺子？

B：大概一周至少两次。

A：那你最喜欢喝什么？

B：我最喜欢的饮料？通常是茶。

A：你每天都喝茶吗？

B：没有。要是喝太多我会失眠的。

Section B

A: What's your hobby, Fangfang?

B: I love playing basketball.

A: Well, I used to think that girls do not like playing sports.

B: You are right, most girls do not like playing sports, but I do.

A: That means you have a healthy body, right?

B: Yes. What about you?

A: Eh... I have a lot of hobbies.

B: How about picking up some to talk about?

A: Sure. I love playing ping-pong, swimming, and I also like playing the guitar.

B: That sounds amazing! You can play the guitar!

A：芳芳，你有什么兴趣爱好？

B：我喜欢打篮球。

A：我以前以为女孩不喜欢体育运动呢！
B：是啊，大部分女孩不喜欢体育运动，但是我喜欢。
A：那说明你的身体一定不错了？
B：是啊。你有什么爱好呢？
A：嗯……我有很多的兴趣爱好。
B：挑选几个说说。
A：我喜欢打乒乓球，游泳，还喜欢弹吉他。
B：太意外了，你还会弹吉他！

Writing

How to Write Product Advertisement

If you are in business and you want to attract customers and make profits, you have to advertise your products. A good advertisement can attract customers' attention, generate their interest in your products, and leave consumers with a strong desire to buy them.

Methods

- Decide where to publish the advertisement.
- Tailor it to your audience.
- Write an attention-grabbing headline.
- Don't start with a question.
- Write a bridge to keep customers reading.
- Create desire for your product.
- Tell customers how to get your product.

【Sample】

TABLE FLOOR ELECTRIC FAN

With remote controller and program controller, Shengmei Brand series fans are equipped with our latest developed integrated circuit-AS9201, which has the most advanced program and the latest formula in the industry.

Shengmei electric fans will provide you with moderate wind of three wind speed selections.

Shengmei electric fans will also provide you with the rhythmic wind once you have a short rest in the day, which reminds you of natural breeze.

When you feel sleepy, Shengmei electric fans will blow a soft and comfortable wind to help you go to sleep. The wind will decrease gradually before it stops automatically at your scheduled time.

You will have 15 selections of time setting and wind speed. The time selections vary from 0.5 hour to 7.5 hours as you want it to be.

Main Specifications

Rated Voltage: 220 V AC. 50Hz

Power: 46 W

Weight: 6 kg

Manufacturer: Shenzhen Advanced Science Machinery and Electronics Corp.

Address: East of Bagua 4 Road, Shenzhen, China

Tel: 0755-226×××××

Assignment: *Write a simulated product advertisement of a harvester.*

_____
_____
_____
_____
_____
_____

Study for Fun

Relax and Enjoy the Song

Directions: *Some words are missing in the song. While you are listening, please fill in the missing words.*

500 miles

If you miss the train I'm on

You will know that I am gone

You can hear the whistle _____

Unit Five

Agricultural Product Processing

A hundred miles

A hundred miles, a hundred miles

A hundred miles, a hundred miles

You can hear the whistle _____

A hundred miles

Lord I'm one, Lord I'm two

Lord I'm three, Lord I'm four

Lord I'm five hundred _____

Away from home

Away from home, away from home

Away from home, away from home

Lord I'm five hundred miles

Away from home

Not a _____ on my back

Not a penny to my name

Lord I can't go back _____

This away

This away, this away

This away, this away

Lord I can't go back _____

This away

If you miss the train I'm on

You will know that I am gone

You can hear the whistle _____

A hundred miles

Unit Six

Plant Protection

Lead-in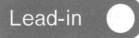

Discussion: What is the Relationship Between Plants and Humans?

● Plants can provide humans with a source of food, building materials, clothing, paper, transportation, etc.
● Plants can provide humans with oxygen.
● Plants can maintain and improve the environment, making it more suitable for humans.
● Some plants have scientific research value and bionics value.
● Plants have economic and aesthetic value.

Intensive Reading

Effects of Elevated Root-Zone CO_2 on Root in Oriental Melon Seedling Roots

Rhizosphere CO_2 is **vital** for crop growth, development, and **productivity**. However, the responses and **effects** of plant mechanisms to root-zone CO_2 are unclear. **Oriental** melons are sensitive to root-zone gas, often encountering high root-zone CO_2 during cultivation. We investigated root growth and nitrogen **metabolism** in oriental melons under T1 (0.5%) and T2

(1.0%) root-zone CO_2 concentrations using **physiological** and comparative analysis. T1 and T2 increased root **vigor** and the nitrogen content in the short term. With increased treatment time and CO_2 concentration, root **inhibition** increased, characterized by decreased root absorption, incomplete root cell structure, accelerated starch **accumulation** and **hydrolysis**, as well as cell aging.

The CO_2 concentration changes continuously with different soil **aeration** conditions, which has a great impact on the growth, development and yield of crops. The CO_2 concentration in the soil close to the plant root system often reaches values up to ten-fold that of the **ambient** atmosphere. Root and soil micro-**organisms** produce CO_2 through **respiration**, which accumulates in the root zone at concentrations normally between 0.2% and 0.5%, but can reach 20% under special circumstances. Responses to high CO_2 soil environment have received increased attention recently in several crop species. However, little information is available regarding the molecular mechanisms of plants in response to **elevated** root-zone CO_2 conditions.

The effects of **excessive** root-zone CO_2 on plant growth, nutrient absorption, and utilization vary with plant species. Nitrogen is an essential **macro**-nutrient for plant growth and basic metabolic processes. High levels of CO_2 in the root-zone promoted the growth of tomato seedlings and increased their NO_3 **uptake**, especially under **salinity** stress and high air temperature; however, there was no significant difference in $NH4^+$. In **lettuce**, high levels of root-zone CO_2 could **alleviate** the midday depression of **photosynthesis** and negative impacts of high air temperature on photosynthesis, and promoted NO_3 uptake and the growth of lettuce plants in the greenhouse. By contrast, high root-zone soil CO_2 had a negative impact on **morphological** and physiological indicators, such as plant height, root length, **chlorophyll** content, photosynthesis rate, **stomata** conductance, NO_3 absorption and **assimilation** in soybean, **maize**, **barley**, and **bean**.

Previous studies have implied that elevated root-zone CO_2 acted as a weak acid, causing **acidification** in root cells, and inhibition of nutrient uptake and the root respiration rate. Moreover, a high soil CO_2 concentration itself might be toxic to plant growth in many plant species, and under certain conditions, CO_2 toxicity is a more important factor in plant growth than CO_2 deficiency.

Thus, elevated CO_2 concentrations in the root-zone could have either positive or negative consequences for plant growth. The differences in the effects of root-zone CO_2 on plants could be caused by differences in plant species, treatment time, the plant developmental period and the CO_2 concentration applied.

The oriental melon is one of main agricultural products that are widely cultivated in some eastern Asian countries. It is sensitive to the root-zone gas environment, and often suffers from root-zone low CO_2 and high CO_2 stress in **irrigated** field cultivation. The responses of melon to root-zone hypoxia have been widely reported. By contrast, there is little information on the mechanism of the oriental melon's response to elevated root-zone CO_2. In addition, the molecular mechanism of the influence of root-zone CO_2 on plant growth and mineral nutrient absorption has not been definitively proved.

Word Bank

rhizosphere ['raizəuˌsfiə]	n.	the soil region of the roots of a plant 根围
vital ['vaitl]	a.	necessary or essential 必不可少的；极重要的
productivity [ˌprɔdʌk'tivəti]	n.	the power to produce 生产力；生产效率
effect [i'fekt]	n.	(on sb./sth.) a change or a result 效应；影响
oriental [ˌɔːri'entl]	a.	related to the eastern part 东方的；东方人的
metabolism [mə'tæblizəm]	n.	process by which food is build up into living matter 新陈代谢
physiological [ˌfiziə'lɔdʒikl]	a.	of the biological study 生理的；生理学的
vigor ['vigə(r)]	n.	active strength of body or mind 活力；精力
inhibition [ˌinhi'biʃn]	n.	the act of restricting or preventing 阻止；抑制
accumulation [əˌkjuːmju'leiʃn]	n.	an increase by growth or addition 积聚；累积

Unit Six
Plant Protection

hydrolysis [haiˈdrɒlisis]	n.	a chemical reaction of water 水解作用
aeration [ɛəˈreiʃn]	n.	the process of being exposed to air 通风
ambient [ˈæmbiənt]	a.	related to the surrounding area 周围的；氛围的
organism [ˈɔːgənizəm]	n.	a living being of microscopic size 细菌；微生物
respiration [ˌrespəˈreiʃn]	n.	the act of breathing 呼吸；呼吸作用
elevate [ˈeliveit]	v.	increase the level of sth. 提高；使升高
excessive [ikˈsesiv]	a.	beyond normal limits 过多的；过度的
macro [ˈmækrəu]	a.	a large quantity of 大量的；最大的
uptake [ˈʌpteik]	n.	the process of taking into the body 吸收；吸入
salinity [səˈlinəti]	n.	the relative proportion of salt in a solution 盐度
lettuce [ˈletis]	n.	a plant with large green leaves 莴苣；生菜
alleviate [əˈliːvieit]	v.	make easier or relived 缓解；减轻
photosynthesis [ˌfətəuˈsinθəsis]	n.	the process of green plants using energy obtained from sunlight 光合作用
morphological [ˌmɔːfəˈlɔdʒikl]	a.	of the morphology 形态学的；形态的
chlorophyll [ˈklɔrəfil]	n.	green pigments found in organisms 叶绿素
stomata [ˈstəumətə]	n.	holes for air vibration 气孔；气洞
assimilation [əˌsiməˈleiʃn]	n.	the state of being absorbed 同化；同化作用
maize [meiz]	n.	a cereal grass bearing kernels on large ears 玉米
barley [ˈbɑːli]	n.	a grain that is used to make food, beer, and whisky 大麦；大麦粒

bean [biːn]	n.	a seed of a climbing plant 豆；豆科植物
acidification [əˌsidifiˈkeiʃn]	n.	the process of being converted into an acid 酸化
irrigate [ˈirigeit]	v.	supply water through pipes or channels 灌溉

Extended Reading

Guest Introduction in Business Meetings

How do you introduce people with status?

The person-of-importance rule applies in most situations. A client may not be an authority, but he or she is certainly important! For instance, you can introduce a colleague to a customer or client："Mr. Zhao, I'd like you to meet my colleague, Ms. Wang." Or you may introduce a non-official person to an official person："Senator Zhao, may I present Mr. Liu?"

If both persons are equal in status, you can fall back on the traditional rules, introducing a younger person to an older person or a man to a woman.

Who is more important?

In business-social situations, such as an official cocktail party which might include spouses, partners, and special guests, business rules should be applied. But who's more important? Is your boss or your spouse? The answer depends on knowing both your boss and your spouse, and gauging the situation accordingly. Perhaps your spouse will understand enough to agree to be slighted. Perhaps your boss is the sort who will appreciate an acknowledgement of the importance of family.

We tend to introduce our boss first in the settings that have been selected by the company, even those outside the office, since our spouses will likely acknowledge the importance of our supervisor in this sphere. If we are at party in our home or if it is a social encounter outside the office, we'd introduce our spouses first.

Should you use first names?

It is always better to call people by their honorary title (Mr./Ms./Mrs.)

and last name until you are asked to use first names or are sure that first names are appropriate. Ms. is usually the preferred title for women. However, the informal rules of your company's corporate culture will apply here. In some places, the chairman of the board is known as "Wang Qiang", while elsewhere, it's always "Mr. Wang". When in doubt, don't use first names.

Exercises

I. Reading Comprehension

Directions: *Choose the best answer for the following questions.*

1. What can be learned from paragraph 2 of the text in the Intensive Reading section? (   )

    A. The mechanisms of plants' responses to root-zone CO_2 are unclear.

    B. For plants to grow normally, a good rhizosphere gas environment is required.

    C. The oriental melon is one of main agricultural products that are widely cultivated in some eastern Asian countries.

    D. Nitrogen is an essential macronutrient for plant growth and basic metabolic processes.

2. What causes the difference in the effects of excessive root-zone CO_2 on plant growth, nutrient absorption and utilization? (   )

    A. Climate.

    B. Water.

    C. Plant species.

    D. Temperature.

3. What is not the effect of elevated root-zone CO_2 acted as a weak acid? (   )

    A. Causing acidification in root cells.

    B. Inhibition of nutrient uptake.

    C. The root respiration rate.

    D. Greenhouse effect.

4. The CO_2 concentration in the soil close to the plant root system often

reaches values up to _____ that of the ambient atmosphere. (　　　)

　　A. once

　　B. twice

　　C. five times

　　D. ten times

5. What can we conclude from the text? (　　　)

　　A. We should protect the environment.

　　B. Plants need sunlight, air and water to grow.

　　C. Rhizosphere CO_2 is vital for the growth, development and productivity of crops.

　　D. The role of fertilizers.

II. Vocabulary

Section A

Directions: *Match the following English words or phrases with their Chinese equivalents.*

1. (　) acid rain		A. 沙漠化
2. (　) forestation		B. 生态系统
3. (　) biosphere		C. 生态灭绝
4. (　) desertification		D. 全球变暖
5. (　) ecocide		E. 酸雨
6. (　) ecosystem		F. 土壤污染
7. (　) global warming		G. 植树造林
8. (　) greenhouse effect		H. 光合作用
9. (　) photosynthesis		I. 温室效应
10. (　) soil pollution		J. 生物圈

Section B

Directions: *Translate the following words and phrases.*

Unit Six
Plant Protection

1. 庄稼
2. 淀粉
3. 大豆
4. 莴苣
5. 叶绿素
6. 二氧化碳
7. 气孔
8. 土壤
9. 微生物
10. 大麦
11. productivity
12. cell
13. root-zone
14. nitrogen
15. plant growth
16. plant height
17. root length
18. weak acid
19. field cultivation
20. absorption

Ⅲ. Situational Speaking
Inviting Guests

Section A

A: Do you have some time tomorrow?
B: Yes, I do.
A: How about having lunch with me?
B: Good idea.
A: What time would be good for you?
B: How about 12:30?
A: Good. I'll see you at 12:30.
B: See you!

A: 明天有空吗?
B: 有啊。
A: 一起吃午饭如何?
B: 好主意。
A: 你觉得什么时间合适呢?
B: 12:30 如何?
A: 好的。我们 12:30 见!
B: 不见不散!

Section B

A: I'm sorry, but I have to cancel out luncheon appointment.
B: I'm sorry to hear that.
A: I have pressing business to attend to.
B: No problem.
A: We'll make it later in the month.
B: Sure.

A: 真抱歉，我不得不取消我们午餐的约会。
B: 太遗憾了。
A: 我有紧急的事情要处理。
B: 没关系。
A: 这个月改天再说吧。
B: 没问题。

Writing

Invitation Letters

An invitation letter is a document that presents a formal request for the presence of an individual, a group of people or an organization at an event. An invitation letter may be formal or informal. It could be printed on paper or sent via email.

Content of Invitation Letters

- Letterhead, including the date, name and address of the inviter, the name and address of the invitee.
- Addressing, including "Sir", "Madam", "Mr.", or "Miss".
- Body.
- Time, place and brief statements.
- Close.
- Genuine sincerity to be invited.

【Sample】

Dear Sir,

We are staging an important exhibition fair of agricultural products in Shanghai Exhibition Centre from October 1st to October 7th.

We would like to invite your corporation to attend. Full details on the exhibition fair will be sent in a week.

We're looking forward to hearing from you soon and hope that you will be able to attend.

            Yours faithfully,
           General manager, ×××

Assignment: *Translate the following simulated invitation letter of business meeting into English.*

邀 请 函

敬启者：

  我们将于2023年5月1日至7日在重庆展览中心举办第十届重庆·中国西部农产品交易会。届时会有1 500家全国知名农产品企业前来参展。交易会以"绿色生态、扩大内需、开放合作、共赢发展"为主题，以企业展示展销为主体，以推动经济和分享商机为宗旨。活动期间，除举办农产品展示交易外，还将举行农产品贸易订单和农业合作项目签约等系列活动。

  诚挚邀请贵公司携产品参加交易会。展览的详细资料将于一月内发出。请尽快回复！

                组委会敬上
                2022年6月1日

Study for Fun

Animal Proverbs in English

Directions: *Match the proverbs and idioms below.*

不入虎穴，焉得虎子	Misfortune might be a blessing in disguise.
如鱼得水	Nothing ventured, nothing gained.
狡兔三窟	Like a duck to water.
不要班门弄斧	The frog in the well knows nothing of the great ocean.
井底之蛙	It is a poor mouse that has only one hole.
人人皆有得意时	Never offer to teach fish to swim.
爱屋及乌	The cat shuts its eyes when stealing cream.
一箭双雕	Every dog has his day.
掩耳盗铃	Love me, love my dog.
塞翁失马，焉知非福	Kill two birds with one stone.

Unit Seven

Animal Husbandry and Veterinary

Lead-in

Discussion: What Are the Restrictive Factors of the Development of Animal Husbandry?

- Production of livestock products enters an age of high cost.
- The problem of quality and safety of livestock products has existed for a long time.
- The situation of animal disease prevention and control remains grim and severe.
- The construction of modern livestock and poultry seed industry system still lags behind.
- Technological support capacity of animal husbandry is still insufficient.
- Protection awareness of animal husbandry is weak.
- Profit of animal husbandry is not satisfactory.

Intensive Reading

Huoyan Geese Hypothalamus Genes Profiles

China has the largest goose production in the world. The goose is well known for its strong adaptability, rapid growth, rich **nutrient** content and low input requirement. The Huoyan goose was famous for its higher laying performance and was listed as one of the nationally protected domestic animals by the Chinese government in the year of 2000. But the problems of variety **degeneration**, especially the decreased number of laid eggs, have been very serious and hinder the goose industry development. Thus, it is urgent to identify the **molecular** mechanisms underlying goose reproductive biology and to improve the laying performance of Huoyan geese.

The hypothalamus plays a central role in controlling poultry reproductive activity. To understand the genes involved in egg laying of Huoyan geese, researchers investigated gene **profiles** in the hypothalamus of laying period and ceased period Huoyan geese using **suppression** subtractive hybridization (SSH, 抑制消减杂交) method.

In addition, some other hormones were found to act as an important regulatory factor in poultry reproduction. Therefore, **minute** differences in hypothalamic functions might affect **reproductive** activities.

Researchers used suppressive subtractive hybridization methods to identify differentially expressed genes in the hypothalamus of laying period and ceased period Huoyan geese. Specifically, the molecular technique of suppressive subtractive hybridization has been **preferentially** used in studies. This method has been used to identify the genes expressed in different reproduction **phases** of poultry. But to the best of our knowledge, gene expression profiles of hypothalamus in geese during different reproduction periods have not been investigated so far.

In the experiment, the Huoyan geese were selected from goose breeding farms and raised according to the program used at this farm. The geese were fed with rice grains and with green grass or water plants as supplements whenever possible. Feed was given during the daytime when the geese were **released** into an open area outside the house. Twenty female geese were killed in January to obtain hypothalamus samples of ceased period group. Another twenty female geese were killed in June to obtain hypothalamus samples of laying period group. All hypothalamus samples were quickly **dissected**, frozen in liquid nitrogen, and stored at $-80℃$ until total **extraction**. Twenty hypothalamus tissues from ceased period group and laying period group were extracted respectively according to the instruction of the manufacturer.

Unit Seven
Animal Husbandry and Veterinary

In summary, two subtracted C-DNA libraries (互补脱氧核糖核酸数据库) were first successfully constructed to show differentially expressed genes in hypothalamus of laying period and ceased period Huoyan geese. These genes were mainly involved in **anatomical** structure development, signal **transduction**, nitrogen **metabolic** process, biosynthetic process, cellular protein **modification** process, cell differentiation, transport, cell **adhesion** and reproduction.

The results described here will contribute to our better understanding of the mechanisms that **underlie** egg laying, and improving the laying performance of Huoyan geese.

These findings provided a new source for **mining** genes related to higher laying performance of Huoyan geese, which **facilitates** our understanding of the reproductive biology of geese.

Word Bank

nutrient ['njuːtriənt]	n.	sth. full of energy and tissue 营养物；滋养物
	a.	of or providing nourishment 营养的；滋养的
degeneration [ˌdidʒenəˈreiʃn]	n.	the process of declining 退化；恶化
molecular [məˈlekjələ]	a.	involving molecules 分子的；由分子组成的
profile [ˈprɔfail]	n.	an analysis or representation 轮廓；剖面
	v.	draw or show the outline 描绘轮廓；构想轮廓
suppression [səˈpreʃn]	n.	the act of suppressing sth. 压抑；抑制
minute [ˈminit]	a.	infinitely or immeasurably small 微小的
reproductive [riːprəˈdʌktiv]	a.	reborn or converted 再生的；生殖的
preferentially [prefəˈrenʃəli]	ad.	as a matter of priority 优先地；优惠地

phase [feiz]	n.	a particular stage in a process 阶段；片段
release [ri'li:s]	v.	set free or unfasten 释放；松开
	n.	instance of surrender 免除；放行
dissect [dai'sekt]	v.	cut up to examine scientifically 解剖；剖析
extraction [ik'strækʃn]	n.	the process of pulling out 提取；取出
anatomical [ænə'tɔmikl]	a.	of the structure of organism 解剖的；解剖学的
transduction [træns'dʌkʃn]	n.	the transfer of genetic material 转导；转换
metabolic [metə'bɔlik]	a.	relating to metabolism 新陈代谢的
modification [mɔdifi'keiʃn]	n.	the act of making sth. different 修正；修改
adhesion [əd'hi:ʒn]	n.	the ability to stick firmly to another 黏附；固定
underlie [ˌʌndə'lai]	v.	lie underneath 位于……之下；是……的基础
mine ['main]	v.	extract or dig from ground 挖掘；开采
facilitate [fə'siliteit]	v.	make easier or likely to happen 促进；使容易

Extended Reading

Call-making Etiquettes

Politeness is as important when speaking over the phone as when talking to people face to face.

Ten etiquettes are generally considered proper as follows:

√ For the caller to give his or her name before asking for the person

Unit Seven

Animal Husbandry and Veterinary

desired.

✓ A person should answer the phone with a pleasant "Hello".

✓ If the person called is not there, the person who answers the phone should tell the caller, "I will have him or her return your call as soon as possible."

✓ When taking a message, a person should cheerfully write down the caller's full name, telephone number, and the time of the call.

✓ Be in good spirits when you're on the phone.

✓ Think and focus clearly when making a call.

✓ Choose the right time to call.

✓ Control your call time. Keep calls short.

✓ Keep the volume moderate.

✓ Mute the phone on special occasions.

On the contrary, ten etiquettes are generally considered improper as follows:

✕ Use mobile phones in restaurants, bars, libraries, driving, meetings, hospital wards, classes, performances, concert halls, cinemas, planes, etc.

✕ Answer the call if a mobile phone suddenly sounds off in public places.

✕ Make a long call on the phone when someone else wants to make a call.

✕ Talk on a mobile phone in a public place for more than three minutes.

✕ Eat or make lazy noises while talking on the phone.

✕ Talk loudly as if no one was watching.

✕ Interfere with others' rest time when there is no emergency.

✕ Make a call before 7 a.m. or after 10 p.m.

✕ Make a call during meals.

✕ Speak swiftly and unclearly.

Exercises

I. Reading Comprehension

Directions: *Choose the best answer for the following questions.*

1. In what aspect does the hypothalamus play a central role? (   )

   A. Hormone.

B. Genes.

C. Profiles.

D. Reproductive activity.

2. It can be inferred that the following poultry animals belong to the periodic seasonal breeders except _____ . (   )

A. geese

B. chickens

C. ducks

D. horses

3. The researchers carry out _____ as an experimental method in order to investigate gene profiles in the laying period and ceased period Huoyan geese? (   )

A. suppressive subtractive hybridization

B. two subtracted C-DNA libraries

C. anatomical structure development

D. goose breeding

4. Which of the following words can replace the word "minute" in the sentence "Therefore, *minute* differences in hypothalamic functions might affect reproductive activities…"? (   )

A. Moment.

B. Miniature.

C. Insignificant.

D. Recorded.

5. Which statement is NOT true according to the text? (   )

A. The research findings can facilitate our understanding of the reproductive biology of geese.

B. The tests screened from the libraries can encode various genes associated with different biological processes.

C. During the experiment, all geese were fed with rice grains and were supplemented with green grass or water plants whenever possible.

D. It is urgent to investigate the molecular mechanisms underlying goose reproductive biology and to improve the laying performance of geese.

Unit Seven

Animal Husbandry and Veterinary

Ⅱ. Vocabulary

Section A

Directions: *Match the following English words or phrases with their Chinese equivalents.*

1. ( ) hypothalamus genes	A. 生殖生物学
2. ( ) domestic animals	B. 喂食谷物
3. ( ) variety degeneration	C. 品种退化
4. ( ) laying eggs period	D. 下丘脑的基因
5. ( ) the goose industry development	E. 产蛋期
6. ( ) reproductive biology	F. 鹅产业发展
7. ( ) suppression hybridization method	G. 家禽
8. ( ) minute differences	H. 脱氧核糖核酸数据库
9. ( ) DNA libraries	I. 细微差异
10. ( ) be fed with rice grains	J. 抑制杂交方法

Section B

Directions: *Translate the following words and phrases.*

1. 无菌的
2. 屠宰场
3. 外展肌
4. 腹腔
5. 鸟结核
6. 鸡胚驯化病
7. 维生素缺乏病
8. 轴索反射
9. 鸡胚疫苗
10. 鲍鱼
11. abdominal dropsy
12. lymph node
13. abduction
14. biosynthetic process
15. abdominal breathing
16. abdominal vein
17. abaxial
18. laying period
19. goose breeding farm
20. abdominal worm

III. Situational Speaking
Making a Telephone Call

Section A

A: Good morning. Wilson Association.

B: This is Li Ming speaking. I'd like to speak to Mr. Wilson.

A: I'm sorry, but Mr. Wilson left here just an hour ago.

B: I've been trying to call him for the last ten minutes, but his line was busy. Will he be back soon?

A: I'm afraid not. He is away for the rest of the day.

B: Do you have any idea where he is now?

A: Presumably at the airport.

B: Is there other way I can reach him?

A: Probably not. He has gone out of this town on business. Perhaps by air now.

B: Do you know when he will be back exactly?

A: I'm sorry. I don't know the exact time.

A：早上好！这里是威尔森公司。

B：我是李明。我想和威尔森先生通话。

A：对不起，威尔森先生一个小时前出去了。

B：我这十分钟一直在给他拨电话，可他的电话一直处于通话中。他会很快回来吗？

A：恐怕不能。他一天都不会回来。

B：你知道他现在在哪里吗？

A：可能在机场吧。

B：还有其他办法联系到他吗？

A：没有。他因公出差一趟。现在可能在飞机上。

B：你知道他什么时候会回来吗？

A：抱歉，我不知道。

Section B

A: May I leave a message?

B: Sure.

A: When he comes back, can you have him call me at this number?

B: No problem.

A: I have a face-to-face business appointment with him next week, but I'm afraid I can't make it.

B: Would you like to make another appointment?

A: Unfortunately, I'm leaving here unexpectedly, and I may be away for two weeks. I'll arrange another appointment in two weeks.

B: I see. I'll tell Mr. Wilson that you've called.

A: Thank you!

A: 我可以留个口信吗?

B: 可以。

A: 他回来后,能不能让他给我这个号码回电话?

B: 没问题。

A: 我和他原定下周有个贸易会晤,但是我计划有变。

B: 您想调整会晤时间吗?

A: 很不巧,我要出差两周。两周后我会再约。

B: 明白了。我会告诉威尔森先生您打电话的事儿。

A: 谢谢你!

Writing

How to Write Personal Resume

A resume advertises a person by summarizing one's educational and professional qualifications as well as character strengths for potential employment. Considered as a brief autobiography, a career history, or even a self appraisal, it introduces one's professional objectives, education, work experience and accomplishments in order to open the door of a formal job interview. A well-written resume will make the job seeker look professional, experienced, well-organized, logical, capable, enthusiastic, and most importantly, prepared for the job.

A personal resume is often organized in two formats: chronological format and functional format.

A resume in the chronological format is easy to read, because the factual information about your education and career, presented in a timeline.

A resume in the functional format concentrates on the skills and qualities you would like to display to the prospective employers. In this type of resume, education and work experience are often listed briefly, but current job skills become a focal part and are described in detail.

Factors of Writing a Resume
- Personal information.
- Job objective.
- Education.
- Work experience.
- Other optional information.
- Personal expectation.

Useful Expressions
- strong analytical / management skills in...
- excellent / in-depth / profound knowledge on...
- ability to... / able to...
- skilled in... /proficient in...
- well experienced in...
- good / in-depth knowledge of...
- highly responsible and dependable...
- build productive relationship with ...
- utilize... methods...
- develop and maintain... relationships...
- work in partnership with...

【Sample】

Personal Resume

Bin Zhao: bin. zhao@163. com　132×××××××

OBJECTIVE

To pursue a position of poultry technician in a major poultry farming enterprise

SUMMARY OF QUALIFICATIONS

Six years of experience as a poultry technician in a leading poultry farming enterprise in the northeast area of China

Specializing in chicken hatcheries and duck cultivation

PROFESSIONAL EXPERIENCE

Research Experience

Conducted research on nutritional poultry feed formula to maximize growth and reproduction

Conducted research in the breeding of silky chicken and the propagation of yellow-feathered broiler

Practical Experience

Assisted in controlled chicken and duck hatching

Operated feeding machine or feed by hand if necessary

Training Experience

Completed and passed all company-mandated training in the past six years

EDUCATION

Bachelor of Veterinary Science in Shenyang Agricultural University (2010-2014)

Master of Veterinary Science in Shenyang Agricultural University (2014-2017)

JOB HISTORY

Dawn Poultry Farming Company (2017-Present)

Assignment: *Write a simulated personal resume to apply for a job in a company of agricultural products.*

Study for Fun

Dialogue Appreciation

Directions: *Some words are missing in the dialogue. While you are listening, please fill in the missing words.*

Jane Eyre (Excerpt)

A: Even good friends may be forced to _____. Let's make the most of what time has left us. Let us sit here in _____. Even though we should be destined never to sit here again.

B: That I never will, sir. You know that. I see the necessity of going, but it's like looking on the necessity of _____.

A: Where do you see that necessity?

B: In your bride.

A: What bride? I have no bride.

B: But you will have!

A: Yes, I will. I will.

B: Do you think I could _____ here to become nothing to you? Do you think because I'm poor and obscure and plain that I'm _____ and heartless? I have as much soul as you and fully as much heart. And if God had _____ me with wealth and beauty, I should have made it as hard for you to leave me as it is now for me to leave you.

A: It is you that I love as my own flesh.

经典台词赏析

简·爱（节选）

A：好朋友也会不得不分离。让我们好好利用剩下的时间。让我们在这儿安安静静坐一会儿，以后再也不会一起坐在这儿了。

B：我决不会，先生。你知道，我看出非离别不可，可这就像看到了非死不可一样。

A：你从哪儿看出非这样不可呢？

B：你的新娘。

A：我的新娘？我没有新娘。

B：但你会有！

A：对，我会。我会。

B：你以为我会留下来，做一个对你来说无足轻重的人吗？你以为，就因为我穷、低微、不美，我就没有心、没有灵魂吗？我也有一颗心，我们的精神是同等的。如果上帝赐予我美貌与财富的话，我也会让你难以离开我，就像我现在难以离开你一样。

A：我爱你就像爱我自己。

Unit Eight

Organic Food

Lead-in

Discussion: What do You Know about Organic Food?

- Organic food is unadulterated food produced without artificial chemicals or pesticides.
- In terms of production technology, chemical synthetic substances or genetically modified materials cannot be used in the production and processing of organic food.
- The taste of organic food is significantly different from conventionally grown food.
- Organic food is different from conventional food in nutrients, pesticide residues, heavy metals, etc.
- The size, color and shape of organic food are also different from those of conventional food.
- Organic food is healthy and safe to some extent.
- Organic food is environmentally friendly because its production meets the strict requirements of ecological environment and conservation of water and soil.
- Organic food emphasizes the use of renewable resources and eliminates the use of chemical fertilizers, pesticides, herbicides, agricultural film, synthetic hormones, etc.

Unit Eight
Organic Food

● The organic food industry is a public sector with a promising future because it communicates the ideas of being healthy, protecting environment, increasing agricultural yields and income, promoting social harmony, etc.

● Growing organic food can avoid wasting resources because the relatively high price, good quality and first-class taste of organic food prevent people from throwing away food easily.

Intensive Reading

Importance of Organic Foods over Conventional Counterparts

The production of **organic** foods, one of the most important branches of **ecological** agriculture, has developed rapidly all over the world. In addition to paying attention to sustainable organic production modes and protection of the environment, people have a strong interest in the quality of organic food. Focusing on the differences in nutrition and safety between organic and **conventional** food, a wide range of research in the world has compared conventional and organic agricultural systems. Much attention has been given to the quality differences between conventional and organic foods.

Many consumers choose organic foods because they are supposed to be more nutritious than their conventional **counterparts** in the market.

One research group compared the **protein composition** of milk produced in both conventional and organic **pastures** in Poland. The results showed that the farming systems had no **significant** effect on protein concentrations of milk, although they did notice that a significant difference existed between organic and conventional milk regarding the composition of protein **fraction** during the late pasturing seasons. Protein in cow milk depends on many factors including season of the year, feeding levels, the **breed** and diet habit of the cattle, their health conditions, stage of **lactation**, and specific genetic characters.

In addition, there are also some studies comparing the quality of organic and conventional crops. No significant **correlations** with starch content were found between potatoes grown in organic and conventional fields.

Whether the nutritional value of organic food is superior to that of conventional food has been a **controversial** issue. Although the existing research data could not draw a **definitive** conclusion, the results showed some **trends**. In the study of carbohydrates and proteins, products from different sources did not show significant differences. However, some published papers indicated that organic foods have shown advantages in fatty acid composition. In addition, organic foods usually contain more vitamins, **trace** elements, and antioxidants than conventional foods. Nevertheless, because many of these reports did not control the **equivalent** soil, water, and fertilizer, etc., further experiments should be conducted to compare the nutrients between organic and conventional foods.

From the chemical safety perspective, **pesticides**, heavy metal residues, and **nitrates** remain at a very low level under the legal limit in most of the organic foods. Furthermore, antimicrobial resistance in organic food is lower than its conventional counterparts. However, many reports indicate that no significant difference was found in mycotoxin content between conventional and organic foods.

In this study, we found that the quality of organic foods was related to various aspects of the food production process, such as planting or breeding environment, maturity or harvest period, package, processing, and storage. Therefore, it is essential to take into account these factors when we study the quality difference of foods under different patterns.

As there are few studies on the role and utilization of nutrients in organic food, future related research can be done from human **epidemiology** and animal testing perspectives. Such research may provide further evidence on whether organic food contains more of certain nutrients than conventional food.

Word Bank

organic [ɔːˈgænik]	a.	(of food) produced or practiced without using artificial chemicals（食品）有机的
ecological [ˌiːkəˈlɔdʒikl]	a.	of plants and living creatures to each other and to their environment 生态的；生态学的

Unit Eight
Organic Food

conventional [kən'venʃənl]	a.	traditional or acceptable 传统的；习惯的
counterpart ['kauntəpɑːt]	n.	a person or thing that has the same position in a different situation 对应的人或事物
protein ['prəutiːn]	n.	a natural substance found in meat, etc. 蛋白质
composition [ˌkɔmpə'ziʃn]	n.	the organized way of different parts 成分；构成
pasture ['pɑːstʃə(r)]	n.	grass land for feeding animals 牧场；牧草地
significant [sig'nifikənt]	a.	large or important enough 重大的；显著的
fraction ['frækʃn]	n.	a component of a mixture 部分
breed [briːd]	n.	a particular type of animal in a controlled way 人工培育的牲畜品种
lactation [læk'teiʃn]	n.	the period following birth during which milk is secreted 哺乳期
correlation [ˌkɔrə'leiʃn]	n.	a connection between two things 相关；关联
controversial [ˌkɔntrə'vəːʃəl]	a.	easily causing argument and debate 有争议的
definitive [di'finətiv]	a.	final and to be the best 最后的；最佳的
trend [trend]	n.	tendency or direction 趋势；走向
trace [treis]	n.	a very small amount of sth. 微量；少许
equivalent [i'kwivələnt]	a.	equal in value, amount 相等的；相同的
pesticide ['pestisaid]	n.	a chemical used to kill pests 杀虫剂
nitrate ['naitreit]	n.	a compound containing nitrogen and oxygen 硝酸盐；硝酸盐类化肥
epidemiology [ˌepidiːmi'ɔlədʒi]	n.	[医] 流行病学

Extended Reading

Business Correspondence Etiquettes

Business correspondence is the communication or exchange of information in a written format during the process of business activities. Business correspondence can take place between organizations, within organizations or between the customers and the organization.

Business Correspondence Types

1. Business letters are the most formal method of communication following specific formats. They are addressed to a particular person or organization. Business letter has its special formats. Formal business letters usually contain 12 components, including letterhead, reference, date, inside address, attention line, salutation, subject line, body, complimentary close, signature, IEC block, and postscript.

2. Email is an informal form of business communication. It is the most widely used written communication form and usually has a conversational style.

【Sample 1】 How to Refuse Cancelling Orders?

Dear Sir,

Thank you for your letter of March 12th. We are very sorry that we cannot accept your request for cancelling the order at the last moment. Since your products are ready for shipment, it will be a huge loss for us if the order is cancelled. Thank you for your understanding and cooperation.

Best regards,

×××

【Sample 2】 How to Confirm Payment of Orders?

Dear Sir,

I have received your mail informing us that the amount of ￥100,000,000 as payment for my invoice No. 1234567 has been transferred to our bank. Thank you very much for your cooperation.

Best regards,

×××

Unit Eight
Organic Food

【Sample 3】 How to Confirm Sample Products Delivery?

Dear Sir,

I am writing to inform you that the samples you requested were sent out by Federal Express today. I have also enclosed a quotation. Please let me know at your earliest conveniences as soon as they arrive. I'm looking forward to your earliest response.

Yours sincerely,

×××

Exercises

I. Reading Comprehension

Directions: *Choose the best answer for the following questions.*

1. In addition to paying attention to sustainable organic production modes and protection of the environment, people have a strong interest in the _____ of organic food. (   )

   A. quantity

   B. price

   C. quality

   D. color

2. Why do many consumers choose organic foods? (   )

   A. Because they are cheaper.

   B. Because they are supposed to be more nutritious than conventional foods in the market.

   C. Because they are delicious.

   D. Because they are easy to get.

3. Whether the nutritional value of organic food is superior to conventional food has been a (an) _____ issue. (   )

   A. undisputed

   B. worthless

   C. controversial

D. boring

4. According to the text, which of the following statements is NOT right? (   )

A. Little attention has been given to the quality differences between conventional and organic foods.

B. We also found that the quality of organic foods was related to various aspects of the food production process.

C. No significant correlations with starch content were found between potatoes grown in organic and conventional fields.

D. From the chemical safety perspective, pesticides, heavy metal residues, and nitrates remain at a very low level under the legal limit in the organic foods.

5. In the future, more research on _____ can be done on organic food. (   )

A. the protein composition of milk

B. feeding levels

C. fatty acid composition

D. human epidemiology and animal testing

Ⅱ. Vocabulary

Section A

Directions: *Match the English words or phrases with their Chinese equivalents.*

1. ( ) organic foods		A. 食品质量
2. ( ) ecological agriculture		B. 有机牧场
3. ( ) food quality		C. 营养价值
4. ( ) organic pastures		D. 重金属残留
5. ( ) diet habit		E. 有机食品
6. ( ) conventional crops		F. 食品加工

Unit Eight
Organic Food

7. (   ) nutritional value   G. 饮食习惯

8. (   ) heavy metal residues   H. 动物试验

9. (   ) animal testing   I. 传统作物

10. (   ) food production   J. 生态农业

Section B

Directions: *Translate the following words and phrases.*

1. 消费者
2. 蛋白质
3. 牧场
4. 营养
5. 脂肪酸
6. 维生素
7. 化肥
8. 收获季节
9. 包装
10. 证据
11. environment
12. focus on
13. agricultural systems
14. take into account
15. health condition
16. research data
17. conventional foods
18. different patterns
19. legal limit
20. human epidemiology

Ⅲ. Situational Speaking
Suggestions and Requests

Section A

A: Good morning, madam. Could I help you?

B: Yes, could you show me some one-piece dress?

A: Sure. We have a large assortment of them. Here they are.

B: Yes. You do have a large range.

A: Yes, here are some examples. What sort of one-piece dress are you looking for?

B: I'm not sure. Could you give me some advice?

A: How do you like this design?

B: Oh, I think it looks too plain. Besides, it is too big.

A：Is this one all right?

B：I think it's out of date.

A：OK. This one may give you every satisfaction. It's just in fashion now.

B：Yeah. It's the very thing I want to buy. Can I try it on?

A：Of course. The fitting room is over there.

（A few minutes later.）

A：Oh. It fits you like a glove.

B：Yes, I like it very much. How much?

A：It is 368 yuan.

B：OK. I'll take it. And here is the money.

A：Thanks. Welcome back again.

A：早上好，太太。您需要帮忙吗？

B：是的，你能拿几件连衣裙给我看看吗？

A：当然。我们有很多款连衣裙。都在这儿。

B：是的。的确有很多可选。

A：是的，这儿是一些样品。您在找哪一类连衣裙呢？

B：我不确定。你能给我提供些建议吗？

A：这款怎么样？

B：哦，我认为它看上去太普通了。另外，这个号也太大了。

A：这款怎么样？

B：我觉得这款过时了。

A：好的。这件肯定让您满意。现在正流行呢。

B：是。这正是我想买的。我可以试穿一下吗？

A：当然。试衣间在那边。

（几分钟后。）

A：这衣服真是太合您的身了。

B：是的，我也很喜欢。多少钱？

A：368元。

B：好，我买了。给你钱。

A：谢谢。欢迎再来。

Writing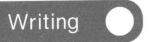

How to Write a Letter of Thanks

Aims
- Express gratitude.
- Be socializing.
- Enhance relationship.

Main Points
- Express your appreciation.
- Highlight the details.
- Express thanks again.

How to Write an Effective Letter
- Acting fast.
- Being short and simple.
- Proof-reading your letter.

Common Words
- appreciate
- convey / express one's appreciation for...
- be grateful to you for...
- express / extend one's gratitude for...
- generosity
- hospitality
- timely help and assistance

First Sentence
- I am writing to extend my sincere gratitude for...
- I am writing to express my thanks for...
- I am writing to show my sincere appreciation for...
- I would like to convey my heartfelt thanks to you for...
- We are grateful to you for...
- I sincerely appreciate...

Details
- Thanks to your effort, we had our most success ever.
- With your help, I made a progress in my study.

- My achievement is owed to your help.
- Owing to your assistance, I have made a progress in...
- Thank you very much for the gift you sent to me.
- It is one of the most wonderful gifts I got on my birthday.

Last Sentence

- I am most grateful for your selfless donation.
- Please accept my gratitude, now and always.
- Thank you for your hospitality and I am looking forward to seeing you soon.

【Sample】

Dear Mr. Wang,

Thank you for your financial support to our technical training program in Shanghai.

The fifteen technical workers who will operate at the newly introduced production line are now under express training of their English proficiencies. They will be sent to your company to receive technical training two months later when their English is good enough for the training in Shanghai.

That's very kind of you to have promised to stand all their expenses in Shanghai. We believe that the training program will guarantee a smooth and fruitful development of our business relations.

Yours sincerely,
Li Ming

Assignment: *Write a thank-you letter to your friend.*

Unit Eight
Organic Food

Study for Fun

Joke Collections

A father of five children came home with a toy, summoned his children and asked which one of them should be given the present. "Who is the most obedient, never talks back to mother and does everything he or she is told?" he inquired. There was silence and then a chorus of voices: "You play with it, Daddy!"

A professor was giving a big test one day to his students. He handed out all of the papers and went back to his desk to wait. Once the test was over, the students all handed in their papers.

The professor noticed that one of the students had attached 100 yuan to his paper with a note saying "One yuan per point."

In the next class, the professor handed the papers back. That student got back his paper and 64 yuan change.

"Xiaoming, what's the matter with your brother?" Mother asked in the kitchen. "He is crying."

"Oh, nothing, Mum," replied Xiaoming. "I'm eating my cake. He is crying because I won't give him any."

"But has he finished'his own cake?"

"Yes." said Xiaoming. "And he also cried when I was helping him finish that."

图书在版编目（CIP）数据

农民实用英语教程 ＝ Practical English Course For Farmers / 姚岚，唐帅主编 . —北京：中国农业出版社，2023.3
农民教育培训农业农村部"十三五"规划教材
ISBN 978-7-109-30548-9

Ⅰ.①农… Ⅱ.①姚…②唐… Ⅲ.①农业－英语－教材 Ⅳ.①S

中国国家版本馆 CIP 数据核字（2023）第 049079 号

NONGMIN SHIYONG YINGYU JIAOCHENG

中国农业出版社出版
地址：北京市朝阳区麦子店街 18 号楼
邮编：100125
责任编辑：郭元建
版式设计：杨　婧　责任校对：周丽芳
印刷：三河市国英印务有限公司
版次：2023 年 3 月第 1 版
印次：2023 年 3 月河北第 1 次印刷
发行：新华书店北京发行所
开本：720mm×960mm　1/16
印张：6.5
字数：122 千字
定价：25.00 元

版权所有·侵权必究
凡购买本社图书，如有印装质量问题，我社负责调换。
服务电话：010-59195115　010-59194918